Science and Its Fabrication

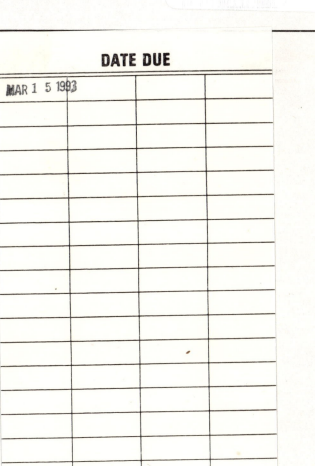

Science and Its Fabrication

Alan Chalmers

University of Minnesota Press
Minneapolis

Published by the University of Minnesota Press
2037 University Avenue Southeast, Minneapolis, MN 55414

Library of Congress Catalog Card Number 90-50276

The University of Minnesota
is an equal-opportunity
educator and employer

Printed in Great Britain

Hugo: I got up early this morning because I've decided to take action. This dawn is the dawn of the unexpected. What's the time?

Joshua: Twelve o'clock, Mr Hugo.

Jean Anouilh, *Ring Around the Moon*

Contents

Contents

Preface

This book is a sequel to *What Is This Thing Called Science?* In that earlier book I subjected some of the standard accounts of science and its methods to a critical scrutiny but did not elaborate on an alternative to them in any detail. I have become convinced that some such elaboration is necessary, especially given the extent to which, against my intentions, my position has been read as a radically sceptical one that denies any distinctive, objective status to scientific knowledge. This book contains an extension and elaboration of the argument of its predecessor. I persist in my rejection of orthodox philosophical construals of the so-called scientific method but show how a qualified defence of science as objective knowledge is possible nevertheless. Consequently, I will no doubt be scorned by many philosophers to my right and sociologists of science to my left.

In a number of places I have drawn on material published in the following articles: 'The Case against a Universal Ahistorical Scientific Method', *Bulletin of Science Technology and Society*, 5 (1985), 555–67 (Chapter 2); 'A Non-Empiricist Account of Experiment', *Methodology and Science*, 17 (1984), 95–114 (Chapters 3 and 5); 'Galileo's Telescopic Observations of Venus and Mars', *British Journal for the Philosophy of Science*, 36 (1985), 175–91 (Chapter 4); 'The Sociology of Knowledge and the Epistemological Status of Science', *Thesis Eleven*, 21 (1988), 82–102 (Chapter 6); 'The Extraordinary Prehistory of the Law of Refraction', *The Australian Physicist*, 12 (1975), 85–8 (Appendix). I gratefully acknowledge the permission of the editors involved to use the material here.

My thanks are also due to Patricia Bower and Veronica Leahy for their efficient and patient typing of the manuscript and to Wal Suchting for helpful criticism.

— 1

The Politics of the Philosophy of Science

1.1 Philosophy of science as a political issue

'In modern times science is highly esteemed.' That is the opening sentence of the book to which this one is a sequel (Chalmers, 1982). Fifteen years of teaching students in a Faculty of Arts, as well as trends in some contemporary philosophy and sociology, have brought home to me the extent to which that assertion needs qualification. Science is commonly seen as dehumanizing, as involving the inappropriate treatment of people and societies, as well as nature, as objects. The alleged neutrality and value-freedom of science is perceived by many as a sham, a perception encouraged by the increasingly common phenomenon of disagreements between experts on opposite sides of a politically sensitive dispute over matters of scientific fact. The destruction and threat of elimination of our environment resulting from technological advances are widely regarded as implicating science. There are those who find the Arts Faculty insufficiently far removed from the oppressive, masculine world of science and turn to mysticism or drugs or contemporary French philosophy. While it surely remains the case that a high regard for science and a generous estimate of its scope form a major component of contemporary ideology, opposing positions are plentiful.

The fact that questions concerning the status of science are politically important has not escaped many philosophers, and more recently sociologists, of science. This is how Imre Lakatos

(1978c, pp. 6–7) summed up the situation in a radio broadcast in 1973.

> The problem of demarcation between science and pseudo-science has grave implications . . . for the institutionalisation of criticism. Copernicus's theory was banned by the Catholic Church in 1616 because it was said to be pseudoscientific. It was taken off the index in 1820 because by that time the Church deemed that the facts had proved it and therefore it became scientific. The Central Committee of the Soviet Communist Party in 1949 declared Mendelian genetics pseudoscientific and had its advocates, like Academician Vavilov, killed in concentration camps; after Vavilov's murder Mendelian genetics was rehabilitated; but the Party's right to decide what is science and publishable and what is pseudoscientific and punishable was upheld. The new liberal Establishment of the West also exercises the right to deny freedom of speech to what it regards as pseudoscientific as we have seen in the debate concerning race and intelligence. All these judgements were inevitably based on some sort of demarcation criterion. This is why the problem of demarcation between science and pseudoscience is not a pseudo-problem of armchair philosophers: it has grave ethical and political implications.

Lakatos, of course, held science in high regard, as did Karl Popper, in whose footsteps Lakatos was keen to follow. Popper (1966, p. 369) explains how his defence of rationality in general and science in particular is an attempt to counter 'intellectual and moral relativism', which he sees as the 'main philosophical malady of our time'. It is not uncommon for defenders of an exhalted status for science to see themselves as defenders of rationality, freedom and the Western way of life for, after all, 'what is really at stake is nothing less than the future progress of our civilization' (Theocharis and Psimopoulos, 1987, p. 597).

Paul Feyerabend is one of the more widely read philosophers who oppose and deride such venerations of science. According to some of his more extreme formulations, current attitudes towards science amount to nothing more than an ideology playing a role akin to that which Christianity played in Western society a few hundred years ago and from which we need to be liberated. Feyerabend (1975) claims that modern science does not possess features that render it superior to and distinct from voodoo or astrology. In his most recent book (1987) he celebrates a 'farewell to reason', where 'reason' is to be read as the mode of rationality presumed to be distinctive of science by those philosophers who defend some privileged status for it. It has become increasingly common in the last few decades for sociologists to turn their attention to the social dimension of science, and especially to the processes involved in the social construction of scientific knowledge. These investigations have led most of them to challenge orthodox accounts of the privileged status of science and some of them to take positions similar to the one defended by Feyerabend. Collins and Cox (1976), for instance, explicitly defend an extreme relativist view and argue that there is no intrinsic difference between the method of science and the method employed by Mrs Marian Keech and her followers to convince others of the authenticity of their dealings with extra-terrestrial beings.

The following pages contain my attempt to illuminate these debates concerning the status of science. A detailed investigation of scientific practice will require us to join Feyerabend and the contemporary sociologists in rejecting much orthodox philosophy of science. However, I will resist the extreme relativism frequently advocated by those authors and will attempt to construct a limited defence of science which extracts what I believe to be correct about the traditional notions of the objectivity and value-freedom of science. That is, I hope that a detailed look at the way in which legitimate scientific knowledge is fabricated (in one sense of 'fabricate') will show how it can be distinguished from its fabrications (in a second sense of 'fabricate'). In the final chapter I indicate why I do not wish my limited defence of the epistemological status of science to be equated with an advocacy of the 'keep politics out of science' kind of attitude, an attitude that leaves the politics that is already in there unchallenged.

1.2 The positivist strategy

The key aim of the logical positivists who flourished in Vienna in the 1920s and 1930s, and whose considerable influence persists, was to defend science and to distinguish it from metaphysical and religious discourse, which most of them dismissed as non-scientific nonsense. They endeavoured to construct a general definition or characteriz-ation of science, including the methods appropriate for its construction and the criteria to be appealed to in its appraisal. With this in hand, they aimed to defend science and challenge pseudo-science by showing how the former conforms to the general characterization and the latter does not. The details of the construal of science offered by the positivists have been rejected or radically modified in recent decades. Nevertheless, the general strategy involved in the positivists' attempt to defend science is still widely adhered to. That is, it is still commonly assumed, among philosophers, scientists and others, that if science is to be defended we require a general account of its methods and standards to appeal to in that defence. Nor were the positivists the first to attempt a general characterization of science. Francis Bacon's *Novum Organum*, René Descartes' *Discourse on Method* and Immanuel Kant's *Critique of Pure Reason* are notable precursors of the endeavours of the positivists to construct a general account of science and its methods.

The general account of science sought by the philosophers I have referred to was intended to be universal and ahistorical. It was to be universal in the sense that it was intended to apply to all claims to science alike. The positivists, for instance, sought a 'unified theory of science' (Hanfling, 1981, ch. 6) which they could employ to defend physics and behaviourist psychology and castigate religion and metaphysics. The account of science sought was to be ahistorical in the sense that it was to apply to past theories as much as to contemporary and future ones. For convenience, I refer to the aim to defend science by appeal to a universal, ahistorical account of its methods and standards as the positivist strategy, since it was such a prominent feature of logical positivism.

Imre Lakatos and Karl Popper are two prominent philosophers of science of recent times who adopt the positivist strategy, although, of course, they are highly critical of the particular account of science offered by the positivists. Imre Lakatos (1978a, pp. 168, 169 and 189) considered the 'central problem in philosophy of science' to be 'the problem of stating *universal* conditions under which a theory is scientific'. He indicated that its solution 'ought to give us guidance as to when the acceptance of a scientific theory is rational or irrational' and hoped that it would help us 'in devising laws for stemming . . . intellectual pollution'. Lakatos appealed to his theory of science to defend contemporary physics and criticize historical materialism and some aspects of contemporary sociology, indicating the universal character he attributed to it, while its ahistorical character is evident from the use Lakatos made of it to argue for the scientific character of the Copernican revolution as well as the Einsteinian one. Alan Musgrave (1974a, p. 560) identifies Popper's cure for relativism as 'an insistence on absolute objective standards'. Popper (1972, p. 39; 1961, section 29) himself sought to demarcate science from non-science in terms of a method that he saw as characteristic of all science, including social science.

It is not uncommon to find working scientists themselves giving expression to the view that a universal account of scientific method can or should be used to defend or help improve science. Thus two contemporary physicists (Theocharis and Psimopoulos, 1987) urge that the practice and defence of science should involve an appeal to an adequate definition of scientific method and deplore the extent to which practising scientists are ignorant of such a definition. They even attribute what they see as the current malaise in science to this ignorance. Other scientists have attempted to settle contemporary controversies concerning the appropriate systems of biological classification by turning to a 'philosophical framework of the criteria of scientific theories and methodologies' (Bock, 1973, p. 381) and see the problem as concerning the 'nature of science' (Gaffney, 1979, p. 80).

The extent to which it is widely and deeply felt that a defence of science must follow the positivist strategy is evident from the typical reaction to those philosophers and sociologists of science who have denied that there is such a thing as a universal, ahistorical account of scientific method and standards capable of guiding the work of scientists or judging the merits of the science they produce. This reaction seems to be motivated by the assumption that giving up the notion of a universal method or set of standards necessarily involves a radical scepticism with respect to science, according to which no scientific theory can be argued to be better than any other, science is epistemologically on a par with astrology and voodoo, and the evaluation of scientific theories is a matter of opinion or taste, an attitude epitomized by the slogan Feyerabend (1975, p. 28) unwisely used to characterize his 'anarchistic' theory of science – 'anything goes'. Theocharis and Psimopoulos (1987, p. 597) are so convinced that a defence of science requires an appeal to a philosophical account of scientific method that they seem to imply that those, like myself, who suggest otherwise to students should be prevented from doing so.

One may wonder how many universities in the world give their science students compulsory formal courses of lectures on the rigours of the scientific method. As for those universities which provide their students with an optional course on the current trends in the philosophy of science, are their governing bodies aware of the fact that many teachers of these courses are bent on sabotaging the scientific method?

In the next chapter I set out my case against the positivist strategy, a strategy which I regard as a very misguided one for those whose aim is to defend science. In subsequent chapters I indicate why the rejection of universal method does not have consequences that need cause any concern to the governing bodies of universities.

1.3   Historically contingent methods and standards

The common, horrified reaction to the abandonment of the idea of a universal, ahistorical method or set of standards, which sees the move as a complete abandonment of rationality, results, I suggest, from a failure to distinguish between the rejection of unchanging, universal method or standards, on the one hand, which I, for one, advocate, and the rejection of all method and standards on the other, which I resist. As I have put it elsewhere (Chalmers, 1986, p. 26): 'There is no universal method. There are no universal standards. But there are historically contingent standards implicit in successful practices. It is not the case that, in epistemological matters, anything goes.' It is not only those who adhere to the positivist strategy that fail to make the distinction between absolute, universal method and standards and contingent method and standards subject to change. Feyerabend (1975, p. 285) is similarly undiscriminating, when, having undermined the orthodox accounts of scientific method, he concludes that 'what remains are aesthetic judgements, judgements of taste, metaphysical prejudices, religious desires, in short, *what remains are our subjective wishes*'.

Can an appeal to contingent standards, such as I advocate, block the way to a kind of sceptical relativism supported at times by Feyerabend and some of the sociologists of science we will discuss later in this book? The fact that an affirmative answer is not straightforward is evident from a common response from those who adopt the positivist strategy to positions such as my own. It is raised, for example, by Barry Gower (1988) in his critique of some of my previously published views. If standards are implicit in successful practices, as I contend, then how can those practices be evaluated from without? More specifically, if, for example, Aristotelian physics embodied Aristotelian standards and Galilean physics embodied Galilean standards, how can we be in a position to argue that Galileo's physics is superior to Aristotle's, as defenders of science would wish to do? If Aristotelian standards are adopted, Aristotle's physics is superior while if Galilean standards are adopted the judgement is reversed. '*Tout comprendre, c'est tout pardonner*' is how Gower (1988, p. 59) sums it up. To argue that Galileo's physics is an advance on Aristotle's do we not need some super-standard applicable to both? And does this not lead us back to the need for a universal method? In similar vein, my opponents can observe that there are methods and standards inherent in astrology or parapsychology and conclude that my position leaves no room for criticizing those practices, since I deny myself recourse to universal standards for evaluating the methods and standards implicit in any practice, however far removed for orthodox science. By following the above line of argument, defenders of the positivist strategy can argue that there is no middle way such as the one I allude to with my talk of contingent standards implicit in successful practices. As for the notion of success referred to here, my critics can insist, as Gower insists, that my use of such a notion is gratuitous unless I have some universal characterization of success. There is no middle way, so this line of argument would seem to suggest. Either we have absolute standards as specified by a universal account of science or we have sceptical relativism, and the choice between evolutionary theory and creation science becomes a matter of taste or faith.

The attempt that I make in this book to capture the middle ground between universal method and sceptical relativism proceeds roughly as follows. In a fairly pragmatic way, and with an eye to what physical science has been able to achieve, I attempt to specify what the *aim* of science is. The aim of physical science is to establish highly general laws and theories applicable to the world. The extent to which those laws and theories are indeed applicable to the world is to be established by pitching them against the world in the most demanding way possible given existing practical techniques. Further, it is understood that the generality and degree of applicability of laws and theories is subject to continual improvement.

Having specified the aim of science in this way, having elaborated it and illustrated it with examples to make it a little less innocuous, and having argued that it is a non-utopian aim frequently fulfilled in science, I am in a position to evaluate methods and standards from the point of view of the extent to which they serve it. How the aim of science is to be evaluated is certainly relative to other aims and interests, but once that aim is adopted, then the extent to which various methods and standards serve it is not a matter of subjective opinion, but a matter of objective fact to be practically established.

Advocates of the positivist strategy typically present themselves as the defenders of science and rationality and their opponents as enemies of science and rationality. In this they are mistaken. In adopting a strategy for defending science that is doomed to failure they play into the hands of the anti-science movement that they fear. They make Paul Feyerabend's job too easy. H. M. Collins (1983, pp. 99–101), a sociologist of science with whom I take issue on a number of occasions in this book, expresses the point I am trying to make admirably, as follows:

> So long as scientific authority is legitimated by reference to inadequate philosophies of science, it is easy for laymen to challenge that authority. It is easy to show that the practice of science in any particular instance does not accord with the canons of its legitimating philosophies. The fears of those who object to relativism on the grounds of its anarchic consequences are being realized, not as a result of relativism, but as a consequence of an over-long reliance on the very philosophies that are supposed to wall about scientific authority. These walls are turning out to be made of straw. If new walls are to be constructed, they will need to have their foundations laid in scientific practice.

I like to think that the defence of science I offer in this book is superior to that of positivist-style defences because it is sustainable, and one that makes clear the terrain on which science is to be defended.

1.4 The critique of pseudoscience

In this book I attempt to portray physical science as an objective and progressive enterprise. The way in which I construct my case involves a close look at what physical science has achieved and how it has achieved it. In particular, my formulation of the aim of science is arrived at in a fairly pragmatic fashion by attending to the kinds of law and theory for the establishment of which adequate methods have been developed in the physical sciences. Because my argument takes this form there are necessary limits to the extent to which my analysis can form a basis for criticizing areas of knowledge outside of physical science. If some area of knowledge, such as Freudian psychology or Marx's historical materialism, to take two favourite targets of philosophers of science, were to be criticized on the grounds that they do not conform to my characterization of physical science, then it would be implied that all genuine knowledge must conform to the methods and standards of physical science. This is not an assumption I am prepared to make, and is one that I think would be very difficult to defend.

One kind of criticism that is possible in the light of my analysis is to challenge claims to knowledge which are presented as if they were scientific in the same sense as the physical sciences, perhaps because they are alleged to have been constructed by similar methods to those of the physical sciences, and which are consequently presented as having a similar epistemological status to that of the physical sciences. If creation science, parapsychology, eugenics or Marian Keech's claims about extra-terrestrial beings (Collins and Cox, 1976) are defended on the grounds that they are scientific in the same sense that physics is, then I believe that the deliberations set out in this book indicate how one might set about repudiating such claims.

When we turn to fields such as social theory or history, which can plausibly be argued to have somewhat different aims and correspondingly different methods and standards to the physical sciences, then my account of science does not have much to offer, nor is it intended to have much to offer, concerning how theories in such fields might be appraised. At best, my analysis and defence of physical science can be taken as indicative of how one might proceed in other cases, that is, by attempting to identify the aims involved, the practices developed to meet those aims and the degree of success attained.

In the penultimate section of *What Is This Thing Called Science?* I summed up my attitude to these issues as follows (Chalmers, 1982, p. 169):

As will by now be clear, my own view is that there is no timeless and universal conception of science and scientific method which can serve the [purpose of appraising all claims to knowledge]. We do not have the resources to arrive at and defend such notions. We cannot legitimately defend or reject items of knowledge because they do or do not conform to some ready-made criterion of scientificity. The going is tougher than that. If, for example, we wish to take an enlightened stand on some version of Marxism, then we will need to investigate what its aims are, the methods employed to achieve those aims, the extent to which those aims have been attained, and the forces or factors that determine its development. We would then be in a position to evaluate the version of Marxism in terms of the desirability of what it aims for, the extent to which its methods enable the aims to be attained, and the interests that it serves.

I hope the discussion in the ensuing chapters will clarify and elaborate on the content of those remarks and show why I feel no need to retract them.

— 2

Against Universal Method

2.1 Introductory remarks

As I indicated in Section 1.2, those who wish to defend some privileged status for scientific knowledge typically adopt what I have termed the positivist strategy. That is, they attempt to define some universal, ahistorical methodology of science which specifies the standards against which putative sciences are to be judged. The influential philosophers of science Popper and Lakatos, while anti-positivist in fundamental respects, adopted a verison of this strategy. More recently, John Worrall (1988, pp. 265 and 274) expresses his allegiance to the positivist strategy quite emphatically. According to him, 'laying down *fixed* principles of scientific theory-appraisal is the only alternative to relativism' so that 'without invariant principles of good science, the whole idea of explaining the development of science as a *rational* process has surely been abandoned'. In similar vein, Barry Gower (1988, p. 59) bemoans the fact that 'the idea of a method characteristic of scientific inquiry is not popular' and attempts to redress the problem.

In this chapter I summarize the reasons why an attempt to defend science by appeal to a universal ahistorical account of science is doomed to failure. Let us suppose, for the sake of argument, that there is a unique category 'science' and a universal scientific method governing its advancement and appraisal. How might philosophers of science establish an adequate characterization of that category 'science' and its method? What resources do philosophers have at their disposal for establishing what science is or should be? I shall look at a variety of candidates for an answer to that question and argue that they are inadequate.

2.2   The appeal to human nature

The attempts by a range of seventeenth-century philosophers to answer my question focused on the importance of human nature. Put in very simple terms, their position can be characterized as follows. Since it is human beings who produce and appraise knowledge in general and scientific knowledge in particular, to understand the ways in which knowledge can be appropriately acquired and appraised we must consider the nature of the individual humans who acquire and appraise it. We must analyse the relevant aspects of human nature. Those aspects are the capacity of humans to reason and the capacity of humans to observe the world by way of the senses. Those who focused on the former were the classical rationalists such as Descartes. Thus, we find Descartes, in his *Discourse on Method* (Anscombe and Geach, 1977, pp. 13–14), having rejected custom and authority as adequate sources of secure foundations for knowledge, resolving to take his studies within himself and to use all the powers of his mind in an attempt to free himself from the 'many errors that may obscure the light of nature in us and make us less capable of hearing reason'. For him, the nature of knowledge, its sources and its limits, were to be understood in terms of our 'natural light of reason'. In the empiricist camp, we find John Locke (1967, p. xxxii) explaining that, faced with some specific epistemological issues, he came to the realization that, before dealing with such issues, it was necessary 'to examine our own abilities, and to see what objects our understandings were, or were not, fitted to deal with'. Very important among these abilities for Locke, of course, was the ability of humans to observe the world by way of the senses. David Hume (1969, p. 42), pursuing the empiricist elements of Locke's epistemology, made it quite clear that, in his opinion, the nature of knowledge is to be understood by investigating the nature of the humans that acquire it. To quote his own words:

> 'Tis evident that all the sciences have a relation, greater or less, to human nature; and that however wide any of them may seem

to run from it, they still return back by one passage or another. Even *Mathematics, Natural Philosophy* and *Natural Religion*, are in some measure dependent on the science of MAN; since they lie under the cognizance of men, and are judged of by their powers and faculties. 'Tis impossible to tell what changes and improvements we might make in these sciences were we thoroughly acquainted with the extent and force of human understanding, and could explain the nature of the ideas we employ, and of the operations we perform in our reasoning.

Rationalist and empiricist theories of science suffer from serious internal problems. The rationalists, when attempting to justify propositions arrived at by clear thinking as true of the world, were, in effect, forced to adopt some problematic notion of self-evidence. (It is worth remembering that most of Descartes' physics, which he attempted to justify by appeal to his rationalist method, turned out to be totally false.) The empiricists were faced with a variety of problems concerning the fallibility and restricted scope of the senses, and with the problem of justifying generalizations that necessarily go beyond the evidence provided by singular applications of the senses (the problem of induction) (Chalmers, 1982, chs 2 and 3). These internal problems are serious, and sufficient to discredit traditional philosophical attempts to found a theory of science on human nature. Nevertheless, I do not consider the internal difficulties faced by traditional rationalism and empiricism to be the main reasons for rejecting them as adequate accounts of science. I suggest that the general approach that involves tracing the nature of scientific knowledge back to the nature of the humans that produce it, is fundamentally misconceived.

Because of the extent to which individual humans are fashioned by the society in which they live, the problem of defining some unchanging essence lying behind the social, cultural and historical differences is notoriously difficult. It is no doubt an essential feature of humans that they are able to think and sense. However, it is not likely to be fruitful to seek the nature of science in whatever is universal in those capacities for the simple reason that, whatever the enduring capacities of humans might be, the reasoning, observational and experimental processes involved in science change and evolve historically. So, for example, the infinitesimal calculus was available for scientists after Newton and Leibniz, but not before, and could be appealed to in defence of talk of infinitesimals in a way that was not available to Archimedes. Again, once Galileo had introduced the technique of testing scientific laws under the artificial conditions of a controlled experiment, it was possible to justify talk of a physical order lying behind the unorderly world of common experience in a way that was not possible before. Galileo's introduction of the telescope opened up a new domain of data to science, and rendered much previous naked-eye data redundant.[1] These facts concerning variations in the reasoning and observational procedures employed in science have nothing much to do with human nature. The differences between the methods of Archimedes and Newton, Aristotle and Galileo, are to be understood not in terms of their respective natures but in terms of the epistemological scenes in which they were immersed. The nature of scientific knowledge, the way in which it is to be justified by appeal to reason and observational procedures, changes historically. To understand and identify it we must analyse the intellectual and practical tools available to scientists in a particular historical context. To attempt to characterize scientific method by looking to human nature is to look in the wrong place entirely.

2.3 The appeal to physics and its history: positivism and falsificationism

Although the traditional approach to understanding knowledge and science, with its focus on human faculties, still has a large influence on orthodox philosophy of science today, a number of contemporary philosophers of science attempt to justify their accounts of science and scientific method in a quite different way. These philosophers accept the message to be drawn from Section 2.2 and conclude that, if we are to understand science and its methods we must focus on science itself and the methods it embodies, rather than on scientists and their nature. Philosophers who adopt this approach typically take physics and its history as exemplifying science at its best. The task of developing an adequate theory of science and its methods then becomes the task of developing a theory that best matches exemplary physics. An account of scientific method is to be tested against the history of physics. Thomas Kuhn, Imre Lakatos and Paul Feyerabend are contemporary philosophers who pay the detailed attention to the history of science implicit in this approach. I shall argue that attempts to justify a universal characterization of science and its method in this way are faced with serious difficulties which undermine the project.

One major difficulty is this. If we demand that an adequate theory of science and its methods be compatible with the history and contemporary practice of physics then we do not have one at our disposal. The main candidates for an adequate account of universal method do not pass the test. This is the main point made by Feyerabend in his book *Against Method* (1975) and it is also one of the main conclusions to which I was led in my previous book (Chalmers, 1982). Here I attempt to summarize the key points made there and elsewhere. Some of my more recent elaborations and extensions of them are included in Chapters 4 and 5.

The positivists aimed to show that legitimate science is 'verified', shown to be true or probably true by reference to 'protocol sentences', facts revealed to careful observers by way of their senses. However, observation statements are public, testable and revisable and quite unlike the positivists' conception of incorrigible truths directly revealed to observers by way of the senses (Chalmers, 1982, ch. 3). The statement 'the Earth is stationary' was accepted as an observable fact for thousands of years before new theories of motion led to its rejection and replacement in the course of the scientific revolution. If we turn to experiment and the role it plays in physics, as opposed to mere observation, then the problem for the positivists' notion that science is based on secure foundations supplied by the senses becomes even greater, as we shall see in Chapter 5.

Even if we concede to the positivists some secure, given, observational base for science, their demand that scientific theories be verified by reference to that base cannot be met. There is inevitably a logical gap between the finite, selective evidence available in support of scientific claims and the generality of those claims themselves. The logical aspects of this point are augmented by the historical observation that many scientific theories of the past, including highly esteemed ones such as Newtonian mechanics, although well supported by a variety of evidence, have been found wanting and have been superseded (Lakatos, 1968). The utopian demands of the positivists have the consequence that our most respected scientific theories are not scientific by their criteria, and even reduce to nonsense for those positivists who hold the view that non-verifiable propositions are indeed nonsense.

The main rival to positivism is Popper's falsificationist account of

science, popular with many practising scientists as well as philosophers. Some of the more general aspects of Popper's position I find
unobjectionable. Scientific theories are fallible and remain subject
to improvement or replacement. In so far as theories make claims
about the world they should be substantiated by pitching them
against the world. The history of science can usefully be understood
as the survival of the fittest theory in the face of severe testing.
However, these concessions to Popper fall short of admitting that
he has successfully pursued the positivist strategy and formulated a
universal, ahistorical account of scientific methodology. If we try to
extract from Popper's writings falsificationist criteria, either for the
acceptance or rejection of theories within a science or for desig-
nating whole areas as scientific or non-scientific, we run into
problems similar to those to which Popper himself has shown
positivism is subject. That is, if we make our falsificationist criteria
too strong then many of our most admired theories within physics
fail to qualify as good science while if we make them too weak few
areas fail to qualify.

Suppose, for instance, we take falsificationism to involve the
demand that falsified theories be rejected. Then, unless 'falsified' is
interpreted so weakly as to be ineffective, exemplary scientific
theories fail to meet the demand. For instance, throughout its
dramatically successful history, Newton's astronomy was confronted by observations incompatible with it, ranging from observations of the moon's orbit to those of the orbit of the planet
Mercury. There are, of course, logical points that render the failure
of scientists to follow our strict falsificationist rule perfectly
understandable and reasonable. Realistic test situations within
science are highly complex. They involve, not only the theory under
test, but a host of auxiliary assumptions, initial conditions and the
like. Matching Newtonian theory with the Moon's orbit involved
asssumptions about the shape of the Moon and its internal motions,
as well as those of the Earth, corrections of telescopic readings to
allow for refraction in the Earth's atmosphere and so on. Eventually
it became possible to save Newton's theory by locating the cause of
apparent falsifications elsewhere in the theoretical maze. As it
transpired, problems posed by Mercury's orbit could not be
disposed of in this kind of way. But it would be most implausible to
expect some falsificationist rule to be up to the task of indicating to
scientists in advance which outcome to expect. It is fortunate that
nineteenth-century physicists were not falsificationists, as defined

by the strict rule under consideration, and that they proceeded to develop Newtonian theory in spite of the unsolved problem of Mercury's orbit. But are we not compelled to be equally charitable, for example, to 'creation scientists' for turning a blind eye to problematic features of the fossil record?

Popper himself does not advocate the strict falsificationist rule discussed above. He recognizes that theories should be given a chance to show their worth, and should not be discarded at the first signs of difficulty. As he puts it himself (1974, p. 55): 'I have always stressed the need for some dogmatism: the dogmatic scientist has an important role to play. If we give in to criticism too easily, we shall never find out where the real power of our theories lies.' Popper's demarcation criterion for distinguishing science from non-science can be split into what we can call a 'logical' and a 'methodological' part. The logical part recognizes that if a theory is to make some substantive claim about the way the world is, then there should be ways in which it can be recognized to be in trouble. That is, there should be possible ways of recognizing that the world is other than the theory claims it to be. This is a reasonable demand that follows from a very general conception of what we understand knowledge of the world to consist in. However, the problem for Popper is that it is satisfied by such a wide range of theories. It was satisfied by Aristotle's physics, for which projectile motion posed a problem. It is satisfied by astrology when some prediction based on it fails to eventuate, and it is satisfied by Freud's theory, since its claim that dreams are wish-fulfilments is threatened by the existence of nightmares and anxiety dreams, to use an example referred to by Popper (1983, section 18) himself. The mere demand for falsifiability, understood simply as the possibility of a clash between the predictions of a theory and some observable outcome, while it might be sufficient to eliminate statements such as 'either it is raining or it is not raining' or some extreme parody of Freudian theory or astrology, admits far more than defenders of the positivist strategy would wish to admit as genuine science.

The second, methodological, aspect of Popper's demarcation criterion is designed to meet the difficulty outlined above. It concerns the character of the appropriate strategy to adopt in the face of apparent falsifications. Theories should be exposed to criticism. They should not be modified in an *ad hoc* way by introducing untestable additions to accommodate problematic evidence. It could be argued that it was in this, unscientific, kind of way that Aristotelians removed the problem posed by projectile motion by introducing untestable assumptions about the motive power of the air through which they moved, while, according to Popper at least, Freud's response to the problem of nightmares was similarly inadequate.

The problem for Popper is that if this aspect of his demarcation criterion is formulated sufficiently strongly to have some force, then physics fails to qualify as a science. Our most prized theories in physics are invariably, and always have been, confronted by problems to which physicists either turn a blind eye or respond in *ad hoc* ways. For instance, in his very first paper presenting the essentials of his kinetic theory of gases in 1859, Maxwell (1965, p. 409) noted that the theory 'could not possibly satisfy the known relation between the two specific heats of all gases'. All of the considerable successes of the kinetic theory occurred *after* the difficulty for the theory was appreciated. It was not removed until the advent of quantum mechanics. Problems in contemporary atomic and nuclear physics are removed by various 'renormalization' techniques widely acknowledged to be *ad hoc*. Why should a highly successful theory with untapped potential be rejected because it faces difficulties which to all appearances can only be accommodated in *ad hoc* ways? What alternatives do modern physicists have but to proceed with the development of the promising aspects of quantum mechanics in spite of any uneasiness they might feel about renormalization? If Popper's falsificationist criterion is given a sufficiently precise formulation to have normative force it has undesirable consequences for science.

The difficulties for Popper's demarcation criterion that I have discussed are precisely those highlighted by Lakatos. His methodology of scientific research programmes was designed to fulfil the positivist strategy by modifying Popper's falsificationism to meet those difficulties. Lakatos's methodology involves a liberalization of Popper's falsificationist criterion. A significant research programme will invariably involve some difficulties, some recalcitrant phenomena, but it need not be abandoned on that count. Evidence conflicting with the central claims of a programme become anomalies rather than falsifications. A programme is scientific if it presents avenues for research, and if this research leads, at least occasionally, to successes in the form of confirmed novel predictions. Anomalies become falsifications of a programme only when it is replaced by a more successful rival which can explain them;

for example, the orbit of Mercury can be said by us, from a post-Einsteinian perspective, to falsify Newtonian theory, while in the nineteenth century it was merely an anomaly.

One problem with Lakatos's demarcation criterion is that it lacks normative force. No research programme can be rejected as falsified because success may be just around the corner, so that 'one may rationally stick to a degenerating programme until it is overtaken by a rival *and even after*' (Lakatos, 1978a, p. 117). Who is to say what significant successes, in the form of dramatically confirmed predictions, lie in store for programmes within contemporary Marxism or sociology, to name two areas unloved by Lakatos. As a tool for combating pseudoscience, Lakatos's methodology is very blunt indeed.

A second major difficulty for Lakatos's methodology stems from the extent to which Lakatos tailored it to match contemporary physics (Feyerabend, 1976). He argues for his methodology by testing it against those episodes in the history of physics over the last two centuries or so which are generally accepted as constituting major scientific achievements (Lakatos, 1978a, p. 124). Given this fact, it becomes quite inappropriate to assume that the demarcation criterion implicit in that methodology is applicable to areas other than physics. Once again, we see that Lakatos's methodology is an ineffective tool for combating pseudoscience.

The above difficulty is one that is faced by all accounts of science and its methods and standards which involve the strategy of attempting to justify general theories of science by appeal to physics and its history. If methods and standards arrived at in this way are presumed to apply generally, to biology, psychology, social theory and the like, what is tacitly assumed is that physics constitutes the paradigm of good science to which all other sciences should aspire. There are prima-facie, and widely acknowledged, reasons to reject this presumption. People, societies and ecological systems are not inanimate objects to be manipulated in the way that the objects of physics can be conceived to be. Artificial experiments and the role they play in physics are typically inappropriate or impossible as adequate means for understanding them. In so far as social and some psychological theories influence people's dispositions and actions, they have an effect on the systems to which they are meant to apply in a way that the physical sciences do not. There is a real sense in which, in developing the human and social sciences, we aim to change rather than merely interpret the world. This is not the place to discuss the special problems that beset social theory, ecology and the like. It is sufficient to note that Lakatos, and those who follow a similar strategy, assume that all legitimate scientific knowledge should share the methods and standards of physics, a position that is difficult to defend and for which Lakatos offers no defence.

2.4   Variable methods and standards in physics

A further difficulty for defenders of universal method and standards emerges once it is recognized that the methods and standards of physics are subject to change, and that they are liable to undergo changes on precisely those occasions when physics makes a dramatic step forward. Scientists transform their methods and standards when they learn, in practice, what is to be gained from such a change. Ironically, an excellent historical example of the point I am making was described in a posthumously published essay by Lakatos (1978b). The argument of that essay poses a serious difficulty for the positivist strategy otherwise defended by Lakatos.

The distinction between science and non-science generally subscribed to in Newton's day was a version of the Ancients' distinction between *episteme* and *doxa*, between genuine knowledge and mere opinion. It was held that genuine scientific knowledge must consist in, or be based on, necessary truths established by reason, while many added the 'essentialist' requirement that those truths be ultimate truths, that is, truths not themselves in need of explanation. Euclidean geometry was often taken to be an exemplary science that lived up to that ideal. Descartes' theory of knowledge, highly influential in Newton's time and regarded by Newton himself as the main account of science to be reckoned with, gave expression to a view of science as based on self-evident, a priori, first principles. Newton's theory clashed with that conception of science. It clashed with the scientific standards of his day. His physics, and especially his account of gravity, could not be proven from self-evident principles. His conception of gravitational action at a distance, far from being self-evident, was widely regarded as unintelligible, a judgement accepted, in a sense, by Newton himself, who acknowledged that while he could describe the action of gravity he could not explain it. Newton's theory did not provide ultimate explanations.

In spite of the clash with the accepted cannons of science, Newton's theory met with dramatic practical success both in astronomy and in terrestrial physics. It was clear that if the fruits of Newton's theory were to be harvested, standards would have to be changed to accommodate it. And that is precisely what happened. Cartesians 'were forced, almost against their will, to oppose the tyranny of self-evident, a priori first principles and thus to change standards of scientific proof and criticism and indeed, the very concept of knowledge' (Lakatos, 1978b, p. 207).

A passage in Lakatos's (1978b, p. 201) essay sums up the situation thus: 'Great works of art may change aesthetic standards – great scientific achievements may change scientific standards. The history of standards is the history of the critical – and not so critical – interaction between standards and achievements.' Provided the analogy with art is not pressed too far, this sums up succinctly my own position. It expresses the fact that standards are subject to change in the light of practical achievements. My analysis of Galileo's introduction of the telescope into astronomy in Chapter 4 provides a further example.

The acknowledgement that standards are subject to change in the light of practice would seem to indicate that the search for a substantive universal, ahistorical methodology is futile, as indeed I think it is. How, then, could Lakatos reconcile his account of Newton's successful transformation of scientific standards with his advocacy of the positivist strategy? I suggest the following quotation provides the clue to what Lakatos's answer would have been (Lakatos, 1978b, p. 220):

> Newton set off the first major scientific research programme in human history; he and his brilliant followers established, *in practice*, the basic features of scientific methodology. *In this sense one may say that Newton's method created modern science.*

The change in method and standards described by Lakatos is interpreted by him as the discovery, in practice, of *the* correct method and standards which presumably was and is to be employed thenceforth in an unchanged form to 'help us in devising laws for stemming . . . intellectual pollution' (Lakatos, 1974, p. 89).

There are two reasons why I regard this position I here attribute to Lakatos as untenable. Firstly, having granted that it is perfectly intelligible to present methods and standards as progressively

changing in the light of practice on one occasion, as Lakatos does
with his study of Newton's physics, it is implausible to assume that
similar changes cannot happen on other, subsequent occasions.
Secondly, it is possible to provide examples of changes in standards
within physics after Newton. For instance, a standard implicit in
nineteenth-century physics involved its determinist character.
Given well-defined initial conditions of a system, its subsequent
development is determined by the laws of physics. The abandonment of strict determinism within quantum mechanics discon-
certed Einstein and others, as is well known. Nevertheless, if we
wish to accept and exploit the practical possibilities for advancement afforded by quantum mechanics we must accommodate
the change in standards involved. The advent of radio astronomy
gave rise to debates concerning what is to count as relevant evidence
in astronomy (Edge and Mulkay, 1976) that are analogous to those
that accompanied Galileo's introduction of the telescope. The
outcome in each case was a significant, progressive change in some
of the standards implicit in experimental astronomy. A third
example is hypothetical, but instructive. Let us suppose, as some
already believe, that reasoning within quantum mechanics does
involve a new 'quantum logic' that violates some classical logical
principles. In such a circumstance the practical success of quantum
mechanics would constitute a good reason for changing our logical
standards in that context. Not even our most hallowed logical
standards are universally given.

   A further conclusion to be drawn from the foregoing consideration reinforces a point made at the end of Section 2.3. If we
acknowledge the extent to which the methods and standards of
physics are fashioned in the light of successful practice we can
recognize the inappropriateness of transferring those methods and
standards to other areas such as sociology or history. Yet this is
precisely what has to be done if the positivist strategy is to be
employed to stem 'intellectual pollution', as Lakatos, for instance,
envisaged.

   In this chapter I have considered two possible answers to the
question: 'What resources do philosophers have at their disposal for
establishing a universal ahistorical account of the scientific
method?' I have considered the appeal to human nature and the
appeal to physics and its history, and I have argued that the question
cannot be adequately answered by such appeals. There is a further
possibility to be considered, one that invokes the aim of science. A

particular methodology of science can perhaps be established on the grounds that it is the one most suited for contributing to the aim of science once that aim is adopted. I consider, and extract what I consider to be of value from, that tactic in the following chapter.

2.5 Note

1 These aspects of Galileo's physics are discussed in some detail in later chapters.

3

The Aim of Science

3.1 Introductory remarks

Although much more needs to be said – and I shall shortly say it – the aim of science can be understood as the production of knowledge of the world, while the aim of the physical sciences, with which I am concerned in this book, can be understood to be the production of knowledge of the physical, as opposed to the human or social, world. In a rough and ready way, at least, it is possible to appreciate the distinction between the aim of, or interest in, producing knowledge and other aims such as serving the economic or political interests of specific individuals, groups or classes.[1] I shall argue, against the sceptics, among whom a number of contemporary sociologists can be included, that techniques have been developed in the physical sciences for producing knowledge that meets the aim of science, suitably interpreted. In the following I attempt to give a rough characterization of the aim of science that serves to distinguish it from other forms of knowledge in a rough way, and then, by attending to the history and practice of physical science, I offer a more substantive characterization of the aims implicit in contemporary science. Methods and standards can be argued for from the point of view of the extent to which they serve a practically realizable version of the aim of science.[2]

Many traditional philosophers approach the problem of analysing science by attempting to develop a general characterization of genuine knowledge and then to understand science as a special case of it (or, as interpreted by the logical positivists, as the only case of it). I have already referred, in the previous chapter, to attempts by the Ancient Greeks to draw a general distinction between genuine knowledge and mere opinion. Near the beginning of the era of modern science we find John Locke (1967, ch. 1, sec. 2) describing his purpose as 'to inquire into the original, certainty and extent of human knowledge, together with grounds and degree of belief, opinion and assent', while David Armstrong (1973) sets out a particularly clear version of attempts by modern analytic philosophers to give a general characterization of knowledge as justified, true belief or something of the kind.

I will not be following any such general approach in my attempt to characterize the aim of science. As the discussion of the previous chapter indicates, I do not regard philosophers as having the resources to be able to formulate a general account of knowledge and its aims which does not involve a careful look at some actual examples of what are regarded as instances of knowledge. Once this is done, it becomes clear, I suggest, that there is such a wide range of kinds of knowledge that the endeavour to find a characterization of knowledge that captures the distinctive features of them all is not destined to be fruitful. Thus, in addition to what is typically regarded as scientific knowledge, we have everyday, common-sense knowledge, we have the knowledge possessed by skilled craftsmen or wise politicians, the knowledge contained in encyclopaedias or stored in the mind of a quiz show expert, and so on. Apart from failing to capture distinctive features of some or all of these various kinds of knowledge, most traditional accounts fail in so far as they are utopian. They specify criteria for genuine knowledge that cannot be satisfied. This is the fate that befalls the various attempts to distinguish knowledge from mere opinion which appeal to notions of necessary or essential truth as characteristic of genuine knowledge.

The immediately preceding comments indicate the way in which I advocate a pragmatic approach to the specification and adoption of aims. If they are to be useful and not futile, aims should not be utopian. They should be such that progress towards achieving them can be made and seen to be made. What is more, whether or not an aim is utopian is something we learn in practice. Our aims can and should be modified in the light of what we learn about what is achievable.

3.2   Science as a quest for generality

A feature of scientific knowledge that I wish to highlight is its generality. If we take uncontentious examples of scientific knowledge, Euclidean geometry and the law of reflection of light, known to the Ancients, say, or Newtonian mechanics and Einstein's relativity theory from more modern times, then the generality of the claims involved is not difficult to appreciate. The theorems of geometry apply in the domains of carpentry, land surveying and astronomy alike, while Newtonian mechanics applies as much to the motion of comets as to the swing of a pendulum.

The importance of generality from a pragmatic point of view is nicely illustrated by Randall Albury's (1983, pp. 44–5) example involving the Dragon's Backbone pump. This pump was used in traditional Chinese society for irrigating rice paddies. Water was carried in pallets which were driven up an inclined square-section trough by a bicycling mechanism. The details of the design of this traditional Chinese pump, especially the shape of the pallets, varied from circumstance to circumstance, presumably as a result of the practical experience of those using it. The pump was introduced into the West in the seventeenth century and used in hydraulic schemes and by firefighting brigades. In the eighteenth century de Belidor, in his *Hydraulic Architecture*, subjected the pump to geometrical and mechanical analysis, and presented a general account of its operation. It is possible, with the aid of de Belidor's analysis, to specify the optimum shape of a pallet for a given circumstance. Whereas the traditional Chinese had possessed craft knowledge based on practical experience, de Belidor's treatment constituted scientific knowledge. The geometry and theory of machines on which it drew were general in the sense that they were applicable to any mechanical situation and the theory of the Dragon's Backbone pump that resulted could be employed to design pumps for novel as well as familiar circumstances.

The foregoing example serves to bring out the connection between generality and utility. While the importance of science as a means of offering improved and extended control over nature has steadily increased since the time of the scientific revolution, many will wish to resist too close an identification between science and its practical application. Science, it will be said, seeks understanding. Improved technology is a byproduct of this improved understanding. Such a view is certainly appropriate in the case of Ancient

Greek and medieval philosophers, many of whom sought to understand the world, the 'reality behind the appearances', with no particular concern with practical applications. Perhaps the same could be said of modern cosmologists, for example. The Ancients sought general knowledge that would explain the everyday world of appearances. For example, taking for granted the observable changes that take place in the everyday world, growth and decay, freezing and boiling, seasonal changes, and so on, they sought an account of the world that would explain how change, in general, is possible. This problem led some of them to propose an atomic theory, by means of which identity through change could be explained in terms of the persistence of the atoms before and after change, while their rearrangement could account for the change itself. According to Democritus, 'in truth nothing exists but atoms and the void'. If anything can be more general than that, then perhaps it is the General Theory of Relativity central to modern cosmology. Whether we view science in terms of the material control or the understanding it affords, generality is one of its distinctive features.

My emphasis on generality needs to be qualified. Important characteristics of science, even contemporary 'pure' science, are lost if we remain too fixated on a picture of science as a quest for theoretical generalities. Ian Hacking (1983) has forcefully illustrated how experiment sometimes, and importantly, 'has a life of its own'. He describes, for example, how David Brewster, a major figure in experimental optics in the first half of the nineteenth century, discovered many new properties of light, thereby providing material that was to be incorporated into the wave theory of light. 'Brewster was not testing or comparing theories at all', Hacking (1983, p. 157) observes. 'He was trying to find out how light behaves.' To give a more modern example, Erwin Hiebert (1988) has described how experimental nuclear physicists were led in a practical way to a 'wave of new experimental findings initiated by the discovery of the neutron, including nuclear fission and self-sustaining chain reactions' which owed little to developments in nuclear theory.

Thomas Kuhn (1977a) has drawn an illuminating distinction between what he calls mathematical and experimental or Baconian science in the seventeenth century. Mathematical science, such as Newtonian mechanics, involved mathematical laws of a high degree of generality, while Baconian science involved practical know-how based on trial-and-error experiment. The latter involved the purposeful investigation of the behaviour of matter in novel situations, 'twisting the lion's tail' as Bacon put it. Much of seventeenth- and eighteenth-century optics comes into this category, as does the line of inquiry that led to the steam engine and the industrial revolution. None of this effective research is adequately understood as a quest for theoretical generality. It owed little to explicitly formulated theory. Baconian science, as a systematic and widespread practice, was a historical novelty in the seventeenth century, and the effectiveness of the strategy was a historical discovery. It remains as a vital component of scientific practice. An important part of the aim of modern science is the extension of the means of practically intervening in and controlling the physical world by systematically twisting the lion's tail.

There are two reasons, I suggest, why the existence and importance of Baconian science does not render my focusing on generality as a distinctive feature of scientific knowledge inappropriate. The first reason involves considerations similar to those illustrated by the history of the Dragon's Backbone pump. How and to what extent can practical effects created and discerned in specific experimental situations be exploited outside those situations? An adequate answer to that question in a particular case requires an adequate theoretical understanding of the situation. The examples of Baconian science cited above bear this out. Drastic improvements in the design of engines became possible in the light of the general theory of thermodynamics that evolved in the nineteenth century, control over nuclear fission was much advanced in the wake of an adequate understanding of binding energies and the like, and Fresnel's wave theory of light opened up practical possibilities that went far beyond what Brewster had been able to achieve. Without wishing to deny the extent and importance of contemporary Baconian science, it is theoretical generalizations that render science distinct from and more powerful than medieval technology.

A second reason for my focus on the theoretical generalizations involved in science is that it is this aspect of science that has been the main target of extreme relativist or sceptical attack, rather than its practical efficaciousness. After all, the claim that science has led us to improved means of practical control over the material world is very difficult to deny in the contemporary world of computers, heart transplants and nuclear power. I am concerned to defend the theoretical aspects of science from mistaken sceptical criticism, thereby creating the space for a more effective critique of science as it is practised in contemporary science. Where sceptical doubts have been raised about the more practical aspects of science, such as the objectivity of experimentation, I will defend it.

If we adopt this view that the aim of science is the establishment of generalizations governing the behaviour of the world, then it is possible to appreciate that there is a fundamental problem to be solved. How are such generalizations to be substantiated? That there is indeed a problem to be solved here is brought home by the reflection that the world around us is complex and messy, so that it is not possible to discern regularities that might constitute scientific generalizations applicable to it. Outside some areas of astronomy and optics, there are no exceptionless regularities to be observed. Even such likely contenders for law-like regularities as 'heavy objects fall straight to the ground' or 'acorns grow into oak trees' are frequently violated in my own garden, the first by the fall of autumn leaves, the second by acorns which fall on stony ground or are damaged by frosts or birds. In Section 3.3 I attempt to illuminate the nature of the problem of how scientific generalizations are to be substantiated by taking a selective look at the history of science and philosophy to discern some of the solutions that have been offered. We will then be in a better position to appreciate the solutions to it implicit in modern science.

3.3 Early attempts to establish theoretical generalizations

How are exceptionless, scientific generalities to be substantiated, given the disorderly nature of the observable world? The philosophy of Plato and Aristotle included responses to this problem. Plato's solution, as usually interpreted, was to assume that our knowledge claims apply with certainty only to an ideal world distinct from the natural world in which we live, so that, for instance, geometry constitutes genuine knowledge of a world of ideal cubes and triangles, and so on, to which the circular and triangular objects of the real world only roughly correspond at best. This move certainly sidesteps the problem I have posed concerning the relationship between the abstract generalizations occurring in scientific knowledge and the disorderly events occurring in the real world, for the latter become irrelevant for Platonic knowledge. Plato's position hardly constitutes a solution to our problem for those who seek knowledge of the actual world, whatever plausibility it may have in mathematics. Aristotle's response to our problem is of more interest. Recognizing the occasional, if not frequent, disparity between the fundamental claims of his theories of nature and commonplace observations, Aristotle qualified claims such as 'heavy objects fall towards the centre of the Earth' and 'olive seeds grow into olive trees' with phrases such as 'for the most part' or 'as a rule' (Barnes, 1975). Secondly, Aristotle distinguished between essential and accidental behaviour and properties so that, for example, the falling of a leaf is essential whereas its fluttering in the breeze is accidental. It is only of the essential that knowledge is possible.

Qualifying generalizations with phrases such as 'for the most part' is an inadequate solution to our problem. While it may be a device that works reasonably well in biology under normal circumstances since, for example, for the most part, olive seeds do grow into olive trees, there are striking counter-examples in other areas. Bearing in mind the typical behaviour of autumn leaves, falling feathers, and so on, it may well be the case that the number of falling objects that do descend vertically towards the centre of the Earth are in a minority. The issue was taken up by a number of medieval authors, influenced, especially, by Thomas Aquinas (Wallace, 1981, pp. 132–5). Their treatment involved an asymmetry between explanation and prediction. It is not possible to predict in advance, for instance, that a particular seed will grow into an olive tree or that a stone, when dropped, will descend vertically. Accidental occurrences, such as the intervention of birds or winds, may prevent things from taking their natural course. However, many medieval peripatetics argued, if a seed does grow into an olive tree, or a stone does fall vertically downwards, then this can be explained by reference to their essence and the natural causes at work. This form of analysis was referred to as reasoning *ex suppositione*. It was extended to the explanation of natural phenomena that occur only rarely, such as lunar eclipses and the rainbow (Wallace, 1974). It cannot be predicted when a rainbow will occur, but if it does, then its cause can be attributed to the refraction and dispersion of the Sun's light by raindrops.

This, then, is a medieval development of one of Aristotle's responses to what I have posed as the problem of the typical lack of agreement between our theories and readily observable events. On the face of it, reasoning *ex suppositione* does avoid the problem. However, a basic difficulty remains. It concerns the method of arrival at the causal explanations of events which, in line with that mode of reasoning, are presumed to have occurred. The difficulty is closely connected with the second of the Aristotelian responses to our problem mentioned above. How are the generalizations governing the behaviour of light that are involved in the explanation of the rainbow known? Precisely what techniques did Aristotel-ianism offer for distinguishing the essential from the accidental? Neither Aristotle nor his medieval successors had anything like an adequate response to this kind of question. For instance, within Aristotle's physics the distinction between essential and accidental motions relies on the notion of an orderly, spherical, earth-centred cosmos, essential motions being those that serve to maintain that order (Clavelin, 1974, pp. 12–21). No systematic way of establishing the existence and character of that order is offered. In the main, it relies on common-sense assumptions of the time, such as the immobility of the Earth and the distinction between the terrestrial and celestial realms. As S. Gaukroger (1978, p. 124) puts it 'the explanatory structure that Aristotle proposes we operate with is incoherent in that explanations of the kind required cannot be given in principle'. Aristotle was an empiricist in so far as he believed that 'it falls to experience to provide the principles of any subject' (*Prior Analytics*, 1, 30, 46a), and yet experience is incapable of leading to knowledge of necessary causes and of enabling the essential to be distinguished from the accidental.

But perhaps by turning to Ancient and medieval philosophers for an answer to our problem we are looking in an inappropriate place. After all, our discussion in the previous chapter indicated that philosophers are still struggling to produce an adequate account of science, and this book would be largely redundant had they been successful. Let us turn to past science itself, rather than to philosophy, to see whether an adequate means for substantiating generalities was implicit in it.

Obvious strong candidates for adequate scientific knowledge established by the Ancient Greeks are Euclid's geometry and Archimedes' statics, the latter consisting of the theory of the balance, centres of gravity and floating bodies. In these sciences, propositions applicable to the world were logically deduced from what, at the time, could plausibly be construed as self-evident first principles or axioms. I need not elaborate on this point in connection with Euclidean geometry. Archimedes' theory of the balance and centres of gravity treated objects as geometrical shapes endowed with weight. These objects could be suspended by weightless threads from rigid balance arms supported by a frictionless pivot. The principles of the theory involved Euclidean geometry, the assumption that bodies tend to move downwards by virtue of their weight, and symmetry considerations which were taken to be self-evident. (For example, it was assumed that if equal weights are suspended from equal balance arms, balance will result because of the symmetry of the situation.) No actual physical situation will precisely conform to those described within Euclidean geometry or Archimedean statics. Nevertheless, if physical situations roughly conform to the Euclidean or Archimedean descriptions, then the prescriptions of those theories of geometry and statics are presumed to be roughly applicable to them. If this point of view is adopted, then it is no more appropriate to test Archimedes' statics by observing the behaviour of actual balances than it is to test Euclidean geometry by measuring and summing the angles of a material triangle. We thus have some kind of account of the relation between theory and experience which proves to be adequate for a wide variety of static physical situations.

While the science of Euclid and Archimedes relied on self-evident first principles, a more empirically orientated path to generality is implicit in Ancient astronomy. Careful observation of the heavens yielded general knowledge in the form of a specification of the observed orbits of the Sun, Moon and planets, knowledge that was adequate for the prediction of eclipses and conjunctions and for forming the basis of workable calendars. The law of reflection of light is a further example of general knowledge established by the Ancients. While some, such as Euclid, attempted to argue for it by appeal to what they regarded as self-evident principles, Ptolemy considered it necessary to test the law by experiment. Ptolemy also suspected that there was a law governing refraction and described experiments designed to establish it, although here he was less successful. (See my somewhat negative appraisal of Ptolemy's experiments in Chalmers, 1975, reprinted as the Appendix to this volume.)

The promise offered by these early successes of the Ancients was not to be sustained. Substantial advances on their contribution to the quest for generally applicable scientific knowledge were not achieved until the scientific revolution. In retrospect, it is possible to see why this was destined to be so. The techniques introduced by the Ancients for establishing generalities applicable to the complex and disorderly phenomena of the real world were adequate to the task only in a very restricted range of circumstances. The search for self-evident physical principles met with limited success only in those areas where the everyday world of common experience offered an adequate basis for the abstraction of principles that could be construed as self-evident. The limited scope and reliability of this procedure becomes evident as soon as the domain of everyday experience is transcended. We now know, for example, that Euclidean geometry is violated on the astronomical scale, while Archimedes' statics would be useless for predicting the behaviour of a balance in a spaceship. The appreciation of these limitations did not emerge until modern times, of course. More significant for our historical story is the fact that, in many areas, principles that could plausibly be regarded as self-evident were totally lacking. This was precisely the problem that confronted Galileo when he attempted to extend Archimedes' techniques from statics to bodies in motion. Common sense or the world of everyday experience does not provide us with self evident principles capable of yielding a law of fall, for example.

As far as the more empirically orientated successes of the Ancients are concerned, we can appreciate that they depended on some highly contingent features of our physical world. Because our solar system happens to consist of a massive Sun accompanied by a few relatively less massive planets which do not interact significantly, the motions of the Earth and planets are sufficiently regular for significant regularities to be discerned by empirical observation. From a modern point of view we could say that the solar system is a very rare example of a convenient experimental set-up that happens to have occurred naturally. The regular behaviour of light rays under a wide variety of common circumstances can also be attributed to contingent features of our world. The interaction between light and gravitational fields is very small, and the wavelength of visible light is sufficiently small to minimize diffraction effects on the macroscopic level.

Given the techniques developed by the Ancients, their success in establishing general scientific knowledge was inevitably confined to a restricted range of special cases.

3.4   Generality and experiment: Galileo

In Galileo's physics we find a novel solution to the problem of how scientific generalizations are to be authenticated. As indicated in the previous section, the main objective of Galileo's physics can be said to have been the extension of techniques that Archimedes had employed in his statics to deal with bodies in motion (Clavelin, 1974; Shea, 1972). Let us see how this led Galileo to introduce a novel role for experiment into science.

In his early works on motion we find Galileo dealing with idealized situations; balances with frictionless pivots, perfect spheres rolling down perfectly flat, frictionless inclined planes, and the like. In those works Galileo indicated that he was aware of the problem of how treatment of such idealized situations relates to systems in the real word, and he warned that 'one who performs an experiment on the subject should not be surprised if the experiment fails' (Galileo, 1960, p. 68). But this amounts to the admission that Galileo's theory cannot be authenticated by appeal to experiment. Once it is further acknowledged that appeal to self-evidence is also inadequate for the purpose, we can see how Galileo had, at this stage, failed to solve our problem.

Galileo's mature physics did contain a qualitative solution. His science of motion involved the claims that all bodies have a natural propensity to move downwards with a uniform acceleration and that horizontal motion is conserved, claims which, when combined, yielded a parabolic trajectory for projectiles. Galileo (1974, p. 223) was aware that these claims were not in general born out by experience.

> Conclusions demonstrated in the abstract are altered in the concrete, and are so falsified that horizontal [motion] is not equable; nor does natural acceleration occur [exactly] in the ratio assumed; nor is the line of the projectile parabolic and so on.

A basic reason why actual motions do not generally correspond to those described in Galileo's theory is the existence of a variety of frictional impediments to motion.

> Considering merely the impediment that the air makes to motions in question here, it will be found to disturb them all in an infinitude of ways, according to the infinitely many ways that

the shapes of the moveables vary, and their heaviness and their speeds.

Because of problems such as these, the fundamentals of Galileo's theory could be put to the test only in experimental situations especially designed for the purpose. The most famous of these were his inclined plane experiments. Galileo tested his claims concerning inertia and free fall by rolling bronze balls 'well-rounded and polished' down a channel in a beam that was as straight as possible. In order to restrict friction to a minimum 'there was glued within [the channel] a piece of vellum, as much smoothed and clean as possible' (Galileo, 1974, p. 169). The motions that served as exemplifications and tests of Galileo's theory are not of the kind that arise of their own accord. For instance, one important sequence of motions investigated by Galileo involved a ball rolling down an inclined plane, being deflected onto a horizontal plane, and then leaving the plane to fall freely (Drake, 1973). It was necessary for Galileo to construct artificial situations especially designed for the purpose of testing his theory which reduced unwanted effects to a minimum. Galileo introduced a variety of techniques for reducing impediments and for dealing with those that remained which have since become a standard part of experimental practice (Koertge, 1977).

The picture of science that best accommodates Galileo's theory of motion can be summarized as follows. Scientific laws and theories describe the tendencies that systems have to behave in particular ways. In actual physical situations these tendencies will be combined in complex ways, so that few regularities will appear at the level of observable events. By experimentally intervening we can attempt to isolate and investigate the individual tendencies and discern the laws governing them. These laws, the evidence for which is gleaned through experimental interventions, are then assumed to apply to the world outside as well as inside experimental situations (Bhaskar, 1978). This is the Galilean solution to the problem of generalization that has become a commonplace in physical science.

Some qualifications are necessary concerning the character of this 'solution'. There is no a priori guarantee that the laws identified in experimental activity continue to apply outside experimental situations. What can be achieved by assuming that they do so is something that has to be learnt in practice. The success that physical science has enjoyed since Galileo is sufficient to confound the extreme sceptic in this regard, but should not be overestimated. While physical science has proved to be extremely effective for dealing with artificially contrived, technological situations, its capability for dealing with the natural world is limited outside some aspects of astronomy. This is exemplified by the notorious unreliability of weather forecasts or, more seriously, by the inadequacies in our appreciation of the environmental impact of our technological intervention in the natural world.

A second necessary qualification concerns the limited extent to which Galileo can be said to have been aware of the implications of his experimental practice. On my interpretation, Galileo transformed the problematic aim for generality in science into a form that was practically achievable to some degree: 'Identify law-like generalities in simple, and if necessary, artificially contrived, situations and assume those generalities to continue to apply in all situations, however complex.' Needless to say, Galileo did not construe his innovations in this way. He remained attracted to the Euclidean or Archimedean ideal and frequently attempted to present his theory of motion as derivable from self-evident principles, a claim that could not plausibly be maintained and which was incompatible with his experimentation (Wisan, 1978, pp. 3–4).

A third qualification that must be added is that Galileo's method of deploying experiments certainly does not provide a method of establishing generalities with certainty. The epistemological implications of Galilean experimentation are discussed in Chapter 5.

3.5   The substitution of growth for certainty

In Section 3.4 we saw how Galileo's physics, in effect, involved a departure from the notion that science should be based on self-evident truths, while in Chapter 2 we saw how Newton's physics similarly involved a departure from the conception of scientific laws as ultimate truths established with certainty. These moves, that set physical science on its modern path, can be summed up by the assertion that modern science has replaced the utopian aim for certainty by the requirement for continual improvement or growth. The demand for growth implies that a good theory should tell us something that we did not know before. The extent to which a theory leads to the successful prediction of qualitatively novel phenomena becomes particularly significant. (The emphasis on growth and novel predictions is one of the characteristics of the philosophies of science of Popper and Lakatos.)

The importance of the kinds of consideration mentioned above emerged as significant in the clash between Cartesians and Newtonians towards the end of the seventeenth century and into the next. The Newtonians argued, with some justification, that Cartesian physics was able to explain only phenomena already known, and even this was achieved only by postulating mechanisms devised in an *ad hoc* way for the purpose. Thus, etherial vortices were devised to account for the known motions of the planets, and streams of right- and left-hand-threaded particles emitted by magnets and flowing through or failing to flow through right- and left-hand-threaded pores in magnetic materials were postulated to account for known magnetic phenomena. By contrast, the Newtonians claimed, again with some justification, that Newtonian mechanics was able, not only to account for known phenomena such as the planetary motions in an uncontrived way, but was also able to predict previously unknown phenomena such as the lack of sphericity of the Earth, the precise way in which the acceleration of gravity varies with distance from the Earth's centre and eventually, and spectacularly, the return of Halley's comet. The recognition that one of the merits of Newton's theory was the degree to which it led to novel discoveries was stressed, for example by H. Pemberton in 1728 in *A View of Sir Isaac Newton's Philosophy* where he noted how it 'led to the knowledge of such things that it would have been reputed no less than madness for any one before they had been discovered even to have conjectured that our faculties ever have reached so far' (Worrall and Currie, 1978a, pp. 212–13). From a contemporary vintage point we are able to add many spectacular examples of successful novel predictions made possible by physical science – one is the radio waves predicted by Maxwell's theory and produced by Hertz, another is the bending of light rays in gravitational fields, predicted by Einstein's General Theory of Relativity and detected by Eddington.

The appropriateness of the emphasis on the growth and improvement of knowledge, and the special significance of novel predictions, is supported by the following general considerations. As I stressed in Section 2.2, individuals do not construct knowledge alone and from scratch. We are all born into an epistemological scene in which there is already much knowledge and various methods for producing, extending and improving it. I do not put this forward as an a priori truth. Extreme empiricists could conceivably have been right when claiming that individuals build up knowledge in otherwise blank minds from the deliverances of the senses, or Descartes could have been right that individuals are able to establish necessary truths by way of the natural light of their reason. However, there is a wealth of evidence, concerning the nature of human perception, language and learning as well as the history of knowledge in general and science in particular, indicating that they are not right. There is no Archimedean point from which to construct and appraise knowledge. We have no alternative but to start where it is at, and to attempt to add to or improve available knowledge by utilizing or improving the methods to hand. New claims to knowledge are to be appraised against a background of what is already known or accepted. That is, they are to be judged by the extent to which they are an improvement on what came before. The ability successfully to predict novel phenomena is surely an important indication of such improvement.

In so far as modern science involves a replacement of the aim for certainty by the aim for improvement or growth, it represents a lowering of the standards that the Ancients had endeavoured to live up to. It represents the substitution of an achievable for a utopian aim. However, the above discussion indicates a sense in which the requirements placed on modern science are more demanding than those of the Ancients. The requirement for continued growth, and especially for qualitative novelty, is not only extremely demanding, it is something that the Ancients could quite reasonably have regarded as utopian. The extent to which, and the ways in which, modern science has been able to grow and to uncover novel phenomena is a practical discovery or achievement that could not have been foreseen.

3.6   The aim of science

In the light of what has been said so far in this chapter, let us consider and summarize what can usefully be said of the aim of science.

Physical science involves the aim to establish generalizations applicable to the physical world. Some means of substantiating those generalizations is necessary. At least since the time of the scientific revolution we have been in a position to appreciate that scientific generalizations (laws and theories) cannot be substantiated a priori and we also have good grounds for accepting that the demand for certainty is utopian. However, the demand that our knowledge be continually transformed, improved and extended is not utopian.

To what extent does this construal of the aim of science provide a substitute for the universal method rejected in the previous chapter and prevent the collapse into some extreme 'anything goes' position? If we are to pursue the aim of science, then some very general prescriptions concerning method and standards can be defended by reference to my characterization of it. We can demand, for example, that candidates for scientific laws and theories should be vindicated by pitching them against the world in a demanding way in an attempt to establish their superiority over rival claims. We can add that, in the physical sciences, such severe testing (to use Popper's apt term) will usually involve artificial experimentation and that the successful prediction of novel phenomena will be especially significant. Any methods or standards more substantive than these fairly bland assertions will need to be worked out in practice within the sciences themselves.

The above assertions, which amount to little more than very rough schematic guidelines or a particular orientation, while falling far short of the substantive methodology many philosophers have devoted lengthy books and articles to, are sufficient to help combat the more extreme forms of relativism and scepticism. In particular, changes in substantive methods, standards and, if you will, paradigms, can be appraised from the point of view of the extent to which they further the aim of the production of improved and more extensive knowledge. I claim that this can be done, and that science can be, and frequently has been and is, practised in a way that predominantly serves the knowledge-producing interest, rather than being subservient to other personal, class and ideological interests. One of the aims of the remainder of the book is to substantiate this against the anarchism of Feyerabend and the relativism of some contemporary sociologists of knowledge. However, as I stress in the final chapter, this falls short of a sanitization of science that renders it immune to a political and social critique. Rather, I hope my analysis clears the way for such a critique.

The attempt I have made to specify the aim of science needs to be qualified to offset some possible misinterpretations of my position.

While I believe that an adequate conception of the aims of science can be employed to defend science from extreme scepticism and makes possible appraisals of claims to knowledge that have weak normative force relative to that aim, I do not wish to be interpreted as regarding the aim of science to be an absolute good necessarily to be ranked above other aims. It might well be argued that the problem of making equitable use of the scientific knowledge that we have is a more pressing problem than the production of more scientific knowledge in contemporary society.

A second qualification involves the recognition that the practice of science and pursuit of its aims in our or any other society are inevitably interwoven with other practices with different aims. To claim, as I do, that it is possible to *distinguish* the aim of science from other aims is not to make the stronger claim that the various activities can be *separated*. I will say a little more about these qualifications in Chapter 8.

3.7 Notes

1 The view developed here has some affinity with Louis Althusser's (1966, ch. 6 and p. 231) understanding of the production of knowledge as analogous to material production. The Althusserian view is clearly articulated and extended in Suchting (1983).
2 Others (Popper, 1979, pp. 191–205; Watkins, 1985; and Laudan, 1984) have appealed to the aim of science to justify their methodologies, although not in the same way, or with the same conception of the aim of science, as I develop here.

— 4

Observation Objectified

4.1 Empiricist assumptions under attack

Many of those who favour the positivist strategy and seek a general characterization of science and its method see it as essential that science be based on secure foundations. They commonly assume that it is our senses that provide those secure foundations. Science is seen as being based on 'objective' facts established by careful use of the senses.

The empiricist assumption concerning the extent to which an objective observational basis for science is available to us has been harshly criticized by philosophers of science in recent decades. They have stressed the non-given, revisable, fallible, 'theory-dependent' character of observation and observation statements. I adopted this line of argument myself in Chapter 3 of *What Is This Thing Called Science?* (Chalmers, 1982). While I still think there is much that is correct in this critique of empiricist assumptions about the foundations of knowledge, I wish to resist a conclusion that is often drawn from it, and which, for example, my students repeatedly draw from it, namely, that observation necessarily is 'subjective', so that observable 'facts' are relative to observers and dependent on their psychology, their history and their culture.

In this chapter I wish to resist the subjectivist, relativist response to the critique of empiricism for which, it would appear, I am partly responsible. I will explore the sense in which observation, as it is deployed in science, is objective, especially when the senses are aided by appropriate instruments. However, my defence of observation will offer no succour to the empiricist who looks to observation to supply secure foundations for knowledge. I will argue, for instance, against the empiricist, that Galileo's introduction of the telescope into astronomy involved a change in the standards governing what was to be considered to be an observable fact, although I will also argue, against the extreme relativist, that Galileo's move was a progressive step forward from the point of view of the aim of science. I regard attempts to undermine empiricist accounts of physics by appealing to subjective aspects of observation to be misplaced. In Chapter 5 I will offer what I regard as a much stronger case against the empiricist idea that secure foundations for science can be provided by the senses, a case that does not appeal to problematic features of perception.

4.2 The theory-dependence of observation

A common line of argument used to contest the empiricist claim that objective facts are 'given' to careful observers by way of the senses is to stress the extent to which the perceptual experiences of individuals are not objectively determined solely by the physical features of what is observed, but are influenced by the expectations and the background, including the theoretical background, of the observer. Thus a layperson, when confronted by an X-ray picture of a chest, may see only ribs surrounded by blotches where a skilled radiologist sees scars and other signs of infection and disease; an experienced microscopist might see cells dividing where James Thurber (1933) can see nothing but a 'nebulous milky substance'. A more specific example comes from the history of geology, and concerns the horizontal formations that resemble roads on the hillsides of Glen Roy in Scotland. The observable facts, as reported by various geologists, differed from one to the other, depending, it would appear, on their theoretical background and past experience. 'The different theories led to different expectations concerning the extent and position of the roads, and duly different findings were reported by the different observers' (Bloor, 1976, p. 21).

These perfectly legitimate reflections on important features of human perception have been employed by philosophers of science to undermine typical empiricist assumptions about the role of observation in science (Hanson, 1958; Kuhn, 1970). It is not difficult to see how this line of reasoning can lead to a thoroughly relativist position. The argument proceeds roughly as follows. Empiricists assume that human perception provides us with objective facts about the world which constitute the foundations of science. However, human perceptions are not objective. They are influenced and shaped in an important way by the subjectivity of observers, by their cultural and theoretical background and by their expectations and point of view. Judgements about what the observable facts are in a particular situation will vary from person to person, from culture to culture, and from theoretical school to theoretical school. Given this relativity of the observable facts, the science based on them is similarly relative to persons, cultures or theoretical schools.

Considerations of the above kind are now common in the philosophy of science and often occur under the heading 'the theory-dependence of observation'. While I endorse many of the points made in those discussions, I regard the emphasis on the subjective or psychological aspects of perception by individual observers as misplaced, for reasons that I will offer shortly, and as playing into the hands of the extreme relativist. The following extended example illustrates this point.

In his study of Galileo's science, designed to support his case against method, Feyerabend (1975) argues that acceptance of the Copernican theory championed by Galileo involved not only a change in theory, but also a change in what was regarded as the empirical facts. Prior to the Copernican revolution science included facts such as 'the Earth is stationary' and 'the motion of a falling stone is straight' whereas after the revolution it was accepted that the Earth spins on its axis and moves bodily around the Sun, while the straight component of the motion of a falling stone is superimposed on the Earth's motion, so that its actual motion is 'mixed straight and circular'. Thus, as Feyerabend (1975, pp. 89 and 187) puts it, the case that Galileo developed to defend the Copernican theory involved 'a change of experience' and 'a partial revision of our observation language' contrary to orthodox empiricist assumptions.

If we look at the details of Feyerabend's construal of this change in the observational base of science, we find that it is attributed to a subjective or psychological change in observers. His argument runs as follows. When considering the description of a situation by an observer, we can abstractly distinguish between the sensations involved, that is, the mental experiences that an observer undergoes when confronting the situation, and the verbal description of the situation which the observer subscribes to in the light of those sensations. Feyerabend insists that, although for the purpose of analysis we can make the distinction between the sensation and the verbal description, the two stages are not separable in practice. An observer does not first have a sensation, when confronting a falling stone, say, and then subsequently interpret that sensation as being indicative of a stone falling vertically downwards. Rather the observer simply sees the stone falling down, and is subsequently willing to accept the statement 'the stone fell straight down'. Feyerabend acknowledges that the splitting of the two aspects of observation, even for the purposes of analysis, is a simplification which has its limitations since our sensations can be influenced by our linguistic expression of them. This reservation aside, we can abide by the distinction and claim that when an observer confronts and describes a situation, he or she automatically makes a connection between sensation and description, between the mental experience and the verbal description accepted on the basis of the sensation. Feyerabend (1975, p. 73) gives the name 'natural interpretations' to those 'mental operations which follow so closely upon the senses' and which constitute the link between having a sensation and accepting a description. Natural interpretations are instilled in us from birth. We acquire them during the process of learning a language, for they enable us to connect language with observable situations. What is more, the natural interpretations embodied in a language and culture at some time will typically have entered it and become part of the observational process many generations previously. Consequently, their nature, and even the fact that they are present, are not readily apparent to the individual.

According to Feyerabend, observations of a falling stone involved a natural interpretation forming an important part of early seventeenth-century common sense that Galileo needed to challenge. It involved the notion of an absolute space defined essentially by the planetary and stellar system with a stationary earth at its centre. Along with this natural interpretation comes the notion of absolute motion in this space. Absolute motion is assumed to have observable effects. In general, the senses faithfully record real motions. An observer imbued with these natural interpretations will automatically take the observed motion of a falling stone to be its 'real' motion in absolute space. The observation of its linear fall clashes with the consequence of the Copernican theory that its motion should be 'mixed straight and circular'. The Copernican theory is refuted from the point of view of early seventeenth-century common sense, and the natural interpretations that are automatically and unconsciously employed by those who have internalized it. After all, 'how could one possibly be unaware of the fact that the falling stone traces a vastly extended trajectory through space!' (Feyerabend, 1975, p. 75).

The details of Feyerabend's account of how Galileo brought about the necessary change in the observation base of science, with which I have taken issue elsewhere (Chalmers, 1986), need not concern us here. What I wish to emphasize is the extent to which Feyerabend construes the change as a change in the subjective experiences of observers. He sees the change as the replacement of one set of natural interpretations by another. Galileo 'insists upon a *critical discussion* to decide which natural interpretations can be kept and which must be replaced' (Feyerabend, 1975, p. 73). 'Galileo's first step, in this joint examination of the Copernican doctrine and of a familiar but hidden natural interpretation, consists therefore in *replacing the latter by a different interpretation*. In other words, *he introduces a new observation language*' (Feyerabend, 1975, pp. 78–9). He thereby 'restores the senses to their position as instruments of exploration' (Feyerabend, 1975, p. 78). From Feyerabend's point of view, then, the testing ground for theories remains observations made by individual observers. Bearing in mind that, for Feyerabend, natural interpretations are 'mental operations that follow so closely upon the senses' and which are 'so firmly connected with their reactions that a separation is difficult to achieve', the replacement of one set of natural interpretations by another involves the replacement of one set of mental operations by another. Prior to Galileo, then, the normal observer, because of his or her cultural background, everyday experiences, language, and so on, is programmed in a way that leads to one particular set of observational experiences and corresponding observation language, whereas the observer that has been subject to the medicine of Galileo's *Dialogue* becomes programmed in a new way that leads to a new set of observational experiences and a new observation language. The change in observation language is located in the individual observer. It is basically a psychological change.

I find Feyerabend's case against empiricism unconvincing. My hunch is that the experiences that twentieth-century observers undergo when observing stones fall, the Sun rise and the stationary Earth differ little from those undergone by seventeenth-century observers. The relevance and significance that a modern physicist would attribute to those experiences are, however, quite different from what was attributed to them by seventeenth-century opponents of the Copernican theory. There is a strong sense in which Galileo transformed the observational basis of science. He did so by introducing instruments such as the telescope, which will be discussed later in this chapter, and he did so by introducing controlled experiment, which was mentioned in the previous chapter and the full implications of which will be elaborated on in the next. But those changes have little to do with changes in natural interpretations that form part of the psychological make-up of individuals. Feyerabend mislocates the change in the observational base of science implicit in Galileo's physics and, as we shall see, underestimates its extent and significance.

4.3   Objective observation as a practical achievement

The fact that perception has subjective and culturally relative elements has not escaped the notice of scientists. It is just because of this commonplace realization that the need to replace mere observation by observation carried out in standardized circumstances following routinized procedures is appreciated. Mere observation is replaced by measurement and controlled experiment. In this way, many of the idiosyncrasies of human perception can be bypassed. Francis Bacon appreciated the point early in the seventeenth century when he wrote:

> whenever I come to a new experiment of any subtlety (though it be in my own opinion certain and approved), I nevertheless subjoin a clear account of the manner in which I made it; that men knowing exactly how each point was made out, may see whether there be any error connected with it, and may arouse themselves to devise proofs more trustworthy and exquisite, if such can be found; and finally, I interpose everywhere admonitions and scruples and cautions, with a religious care to eject, repress, and as it were exorcise every kind of phantasm (Burtt, 1967, p. 21).

The point that I presume Bacon is making here, and which I endorse, can be illustrated by the following example. The so-called Moon illusion is a commonly experienced phenomenon. The Moon appears much larger in diameter when it is near the horizon than when it is high in the sky. Normal perception, if taken as a reliable guide to the Moon's size, is illusory. But we can do better than rely on the unaided senses. We can, for example, mount a sighting tube fitted with cross-wires in such a way that its orientation can be read on a scale. The angle subtended by the moon at the place of sighting can be determined by aligning the cross-wires with each side of the moon in turn and noting the difference in the corresponding scale readings. This can be done when the Moon is high in the sky and repeated when it is near the horizon. The near identity of the result in the two cases indicates that the angular size of the Moon remains unchanged. Normal perception is indeed, in this case, illusory.

Those who wish to stress the 'theory-dependence' of observation will be quick to point to the 'theory' involved in my method for observing the Moon's size. They could observe, quite correctly, that the significance attributed to the alignment of the sighting tube involves an assumption that can loosely be stated as 'light travels in straight lines', and that the adequacy of the observation of the Moon's size yielded by my method is dependent on the adequacy of this, and other, underlying assumptions. Their point could be strengthened by noting that if a sighting tube of the kind I describe were used to establish the direction in which a star closely aligned with the Sun lay, it would yield incorrect data because in this circumstance light from the star is deflected by the gravitational field of the Sun.

I have no quarrel with observations such as these. I do not wish to deny that the adequacy and significance of observation statements depend on theoretical assumptions of various kinds and are consequently fallible and revisable. What I wish to illustrate with my example is the point that the lack of a secure observational base is not due primarily to the vicissitudes of perception. Science has developed powerful techniques for circumventing those problems. In so far as we can arrange to test scientific theories by way of standardized procedures involving observation of such things as pointer readings, clicks from a counter and computer printouts, the problems stemming from the subjective character of human perception can be minimized. The relevant observations become objectified. The task of arguing that a statement like 'the pointer lies between the 2 and the 3 on the scale' is theory-dependent and fallible is best left to the suitably addicted philosophers. The reasons for rejecting the claim that science has a secure observational base lie elsewhere.

A feature of perception that empiricists tend to overlook, and which scientists exploit, is the extent to which it involves an active engagement with the world rather than a passive contemplation of it. Even in ordinary, everyday, perception we check for the reality of a sighted object, for example, by touching it or by moving the head to see if the image responds in the appropriate way. Popper (1972, ch. 5) has noted this aspect of perception and points out that what is unproblematic about mundane observation statements is not that their truth is revealed to unprejudiced observers via the senses, but that they are able to withstand a variety of simple tests. A microscopist views a red blood cell through an electron microscope and sees a configuration of dense bodies. Do they correspond to structures in the cell or are they artefacts of the microscope? The cell is mounted on a microscopic grid whose squares are labelled. Viewed by means of the electron microscope the dense bodies are seen and their location on the grid noted. The same sample of cell on grid is then viewed by means of a fluorescent microscope involving totally different physical principles than those involved in the operation of the electron microscope. The same arrangement of dense bodies is observed at the same locations on the grid. Can there be any serious doubt that the observed structures (whatever they might be) are indeed present in the cell? (Hacking, 1983, ch. 11). It is the results of our practical interventions that lend objectivity and credence to observation reports.

A view that has support among philosophers of science but which I reject portrays the objective facts on which science depends as those observation statements that normal observers can readily agree upon in the light of the evidence of their senses. This consensus view of observation statements overlooks the importance of skill and training for observation within science. The skilled radiologist can see signs of infection in an X-ray picture and the skilled microscopist can see cells dividing when the majority of observers, lacking the appropriate skills, cannot. If we think of (provisionally) acceptable observation statements as those which have survived the severest tests that can be levelled at them, then one way of severely testing a claim about what is to be observed on a microscope slide is to ask a skilled microscopist to take a look, rather than seeking James Thurber's opinion. Nor is the acceptability of an observation statement to be attributed to the mere fact that the appropriate experts agree. It is the extent to which the statement is able to withstand objective tests that is fundamental. The diagnoses of expert radiologists can be wrong and they can be tested in independent ways, for example by seeking additional symptoms for an alleged infection, or by confronting the infected area directly through surgery.

Acceptable observation statements can be understood as those statements describing observable states of affairs that are able to survive tests involving skilled use of the senses. I appreciate that judgements about the significance and severity of tests, and the range of tests available, will be theory-dependent in various respects so that observation statements will be fallible in various degrees. (We can imagine Ptolemy severely testing, and substantiating, the claim that the Earth is stationary by jumping in the air to see if the Earth moved away beneath him.) But I am arguing for the objectivity of observation in science not for its infallibility.

My insistence that physical science involves observation that is objective in the kind of way I have attempted to characterize is subject to an important qualification. *Objectivity is a practical achievement.* While I claim that it can be achieved, and often is achieved, in physical science there is no guarantee that this will prove possible in all cases. The French physicist, Blondlot, claimed to have discovered a new kind of radiation (N-rays) and published detailed instructions concerning how they were to be produced and observed. He and his associates claimed to see the variations in brightness on a screen that constituted the evidence for them. However, investigators from outside the laboratory were unable to see what Blondlot insisted he could see. Blondlot insisted that his critics lacked the appropriate skill. More to the point, Blondlot's claims failed to survive independent tests. For example, when the American physicist, R. B. Wood, removed the prism that was supposed to be influential in the production of the N-rays, Blondlot, unaware of Wood's action, continued to observe signs of N-rays on the screen. Mary Tiles (1984, p. 60) sums up the situation admirably:

> To insist that experimental observation requires the development of special observational skills, which not everyone may

be able to acquire, is, in itself, correct. Problems occur only when [as in the case of Blondlot] all attempts at indirect, instrumental confirmation fail, so that the only evidence is perceptual and hence highly dependent on the 'sensitivity' of the individual observer. In this situation the phenomenon is left irremediably subjective.

Objectivity is a practical achievement, an achievement that is frequently, though not without difficulty, accomplished in physical science. I should add that the extent to which my account of objectivity is applicable in other areas is something that I leave entirely open. I am not at all sure whether, how, and to what extent objectivity can be achieved, for example, by Western anthropologists investigating an alien tribe. I have no special competence to tackle such difficult but important questions.

In the remainder of this chapter I illustrate and develop my views on observation in science by a detailed example involving Galileo's use of the telescope.

4.4   The significance and problematic character of Galileo's telescopic data

The story of Galileo's introduction of telescopic data into astronomy, as I interpret it, is a story of Galileo's successful struggle to objectify and vindicate such data. My version can be instructively contrasted with Feyerabend's, which he uses to lend support to his anarchistic account of science. According to Feyerabend, the reliability of Galileo's telescopic observations and the Copernican theory they lent support to were refuted by experience and Galileo exploited the harmony between these two refuted ideas to gain support for each of them. He thereby furthered the Copernican cause 'by *ad hoc* hypotheses and clever techniques of persuasion' (Feyerabend, 1975, p. 143). While I will argue that these excesses of Feyerabend can and should be avoided we will nevertheless see that Galileo's move involved a transformation in the observational base of astronomy and a change in the standards governing what is to count as appropriate evidence in science.

In a three-month period, from December 1609 to February 1610, Galileo turned the telescope that he had constructed to the heavens. What he observed had dramatic implications for astronomy and, in particular, for the defence of the Copernican theory. He hastened to publish *The Starry Messenger* (Galileo, 1957, pp. 27–58), in which he reported these early findings, and became an international celebrity as a result.

If taken at their face value, these early revelations of the telescope helped the Copernican cause, although not to the extent that Galileo sometimes implied. For instance, the Earth-like appearance of the mountains and craters on the moon posed a problem for the Aristotelian distinction between the incorruptible, ethereal celestial region, which was presumed to include the Moon, and the changeable, corruptible terrestrial region. The moons of Jupiter served to diffuse an Aristotelian objection to the Copernican theory. According to that objection, the bodily motion around the Sun attributed to the Earth by Copernicus rendered the fact that the Moon remained with the Earth inexplicable. But since the Aristotelians themselves accepted that Jupiter moved, its moons posed a similar problem for them. During the years following his initial observations Galileo made further significant observations. He found that the apparent size of Mars and Venus, as viewed through the telescope, varied in accordance with the predictions of the Copernican theory, as opposed to naked-eye observations which showed little change in apparent size. These observations were the main focus for Feyerabend's charge that Galileo's telescopic data were defended in an *ad hoc* way, as we shall see. Galileo's telescope revealed the phases of Venus and showed them to wax and wane in the way Copernicus had predicted.

However, employing telescopic data to support the Copernican theory begs the important question of why the telescopic data should be accepted in preference to the corresponding naked-eye data. Feyerabend is right to urge the fundamental importance of that question. Why should the evidence revealed by sightings through Galileo's tube fitted with a convex and concave lens be preferred to evidence obtained by the eyes directly?

We first note, along with Feyerabend, that Galileo was not in possession of a theory of the telescope, and his attempt to offer one, when challenged, was blatantly inadequate (Galileo, 1957, pp. 245–6). This circumstance need not be taken as posing too serious a problem for Galileo. The fact that lenses refract light and that individual lenses can magnify was well known, and had been exploited since the late thirteenth century for the manufacture of spectacles. It was not such a great step to presume that a combination of two lenses might do the job better. Secondly, the need to support observations by explicit appeal to theory could be questioned. It could be pointed out that confidence in naked-eye data does not result from an appeal to a theory of the workings of the eye. Let us turn, then, to possible justifications that appeal to practice.

The veracity of telescopic observations of terrestrial objects can be demonstrated in a fairly direct way by virtue of the fact that telescopic data can be checked by close-up unaided observation of the object viewed. Further, our familarity with terrestrial scenes enable us to use a number of visual cues, either consciously or unconsciously, when viewing a particular scene. Thus, for instance, overlap gives a guide to relative distance and size can be estimated by comparison with objects of known size. When we recall that Galileo's telescopes were prototypes made by trial and error using hand-ground lenses we can appreciate the extent to which they must have produced aberrations. If the objects sighted are familiar ones then it is easy for the observer to pick them out from the hazy accompaniments produced by the telescope, and, for example, to ignore the curvature and red and blue colouring exhibited by the image of a distant ship's mast.

When the telescope was turned to the relatively unknown skies, however, these aids to perception were, in the main, lacking. There is evidence of the difficulty in Galileo's own reports. The largest of the craters shown on Galileo's drawing of the Moon cannot be seen with a modern telescope, nor can it be seen if one goes there. It is possible that Galileo's telescope was responsible for that crater, as Feyerabend suggests. Galileo acknowledged that his telescope magnified the stars much less than it magnified the planets, but could not account for this inconsistency. There were real problems facing Galileo concerning the veracity of his telescopic data.

A further obstacle in the path of acceptance of telescopic data was a philosophical view of sense perception that stemmed from Aristotle and was accepted by many of Galileo's opponents. According to that view, the senses necessarily yield reliable information about the world when used carefully under normal conditions. Galileo's biographer, Ludovico Geymonat (1965, p. 45), refers to the belief 'shared by most scholars of the time' that 'only direct vision had the power to grasp actual reality' while Scipio Chiaramonti, one of Galileo's opponents, referred to the view that 'the senses and experience should be our guide in philosophising' as 'the criterion of science itself' (Galileo, 1967, p. 248). Maurice Clavelin (1974, p. 384), in a context in which he is comparing Galilean and Aristotelian science, observes that 'the chief maxim of Peripetetic physics was never to oppose the evidence of the senses', while Stephen Gaukroger (1978, p. 92), in a similar context, writes of 'a fundamental and exclusive reliance on sense-perception in Aristotle's work'.

A teleological defence of the reliability of the senses was common. The *function* of the senses was understood to be to provide us with information about the world. Thus, although the senses can mislead us in abnormal circumstances, for instance in a mist, or when the observer is sick or drunk, it makes no sense to assume that they can be systematically misleading when they are fulfilling the task for which they are intended. Irving Block (1961, p. 9), in an illuminating article on Aristotle's theory of sense perception, characterizes Aristotle's view as follows:

> Nature made everything for a purpose, and the purpose of man is to understand Nature through science. Thus it would have been a contradiction for Nature to have fashioned man and his organs in such a way that all knowledge and science must, from its inception, be false.

Aristotle's view was echoed by Thomas Aquinas many centuries later:

> sense perception is always truthful with respect to its proper objects, . . . for natural powers do not, as a general rule, fail in the activities proper to them, and if they do fail, this is due to some derangement or other. Thus, only in a minority of cases do the senses judge inaccurately of their proper objects, and then only through some organic defect, e.g. when people sick with fever taste sweet things as bitter because their tongues are ill-disposed (Block, 1961, p. 7).

Galileo's introduction of the telescope into science ran counter to the reliance on unaided sense perception, with its teleological underpinning, and Galileo's contemporaries might well have responded to it with the remark that 'if God had meant man to use such a contrivance in acquiring knowledge, he would have endowed men with telescopic eyes', as Kuhn (1959, p. 226) suggests. If Galileo was to be successful in achieving acceptance for his telescopic data it was necessary for him to violate and change 'the criterion of science itself'. Let us see how he succeeded in doing so.

4.5 Galileo's observations of Jupiter's moons

In Section 4.3 I noted that the typical scientific response to the vicissitudes of perception is to attempt to replace mere observation by measurement involving routine procedures under standardized conditions. Galileo's observation of the moons of Jupiter provides an excellent example of such a move.

Galileo soon appreciated the need to fix the telescope on a stable mounting. He also found images to be more distinct if the light entering the telescope by way of the convex lens was restricted to the central portions of that lens by a covering (Drake, 1978, p. 147). When, in January 1610, Galileo first sighted the 'starlets' accompanying Jupiter across the sky, qualitative features of their positions on successive nights led him to believe that they were satellites of Jupiter. Within two years Galileo had devised an objective procedure for measuring the separation of the satellites from Jupiter and this enabled him to mount an extremely strong case for the veracity of the telescopic sightings of the satellites and for the orbits he attributed to them. Galileo's procedure is worth describing in a little detail (Drake, 1983, pp. 128ff.).

A scale was attached to the telescope by a ring in such a way that the plane of the scale was perpendicular to the axis of the telescope and could be slid up and down along its length. A viewer, looking through the telescope with one eye, could view the scale with the other. Sighting of the scale was facilitated by illuminating it with a small lamp. With the telescope trained on Jupiter, the scale was slid along the telescope until the image of Jupiter, viewed through the telescope with one eye, lay between the two central marks on the scale as viewed through the other eye. With this accomplished, the position of a satellite viewed through the telescope could be read on the scale, the reading corresponding to its distance from Jupiter in multiples of the diameter of Jupiter. The diameter of Jupiter was a convenient unit, since employing it as a standard automatically allowed for the fact that its apparent diameter as viewed from Earth varies as that planet approaches and recedes from the Earth. Where necessary, Galileo could transform his relative measurement into absolute measurements of angle subtended at the eye by dividing the angles subtended by the images on the scale by the magnification of the telescope. Galileo had devised a method for measuring the magnification of his telescopes soon after he commenced to use them, and described his method in *The Starry Messenger*.

Using the procedures I have described, Galileo was able to record the daily histories of the four 'starlets' accompanying Jupiter. He was able to show that the data were consistent with the assumption that the starlets were indeed satellites orbiting Jupiter with constant periods. The assumption was born out, not only by the quantitative measurements, but also by the more qualitative observation that the satellites occasionally disappeared from view as they passed behind or in front of the parent planet or moved into its shadow.

Galileo was in a strong position to argue for the veracity of his observations of Jupiter's moon, in spite of the fact that they were invisible to the naked eye. He could, and did, argue against the suggestion that they were an illusion produced by the telescope by pointing out that that suggestion made it difficult to explain why the satellites appeared near Jupiter and nowhere else. Galileo could also appeal to the consistency and repeatability of his measurements and their compatibility with the assumption that the satellites orbit Jupiter with a constant period. Galileo's quantitative data were verified by independent observers, including observers at the Collegio Romano and the Court of the Pope in Rome. What is more, Galileo was able to predict further positions of the satellites and the occurrence of transits and eclipses, and these too were confirmed by himself and independent observers (Drake, 1978, pp. 175–6 and 236–7).

The veracity of the telescopic sightings was soon accepted by those of Galileo's contemporaries who were competent observers, even by those who had initially opposed him. It is true that some observers could never manage to discern the satellites, but I suggest that this is of no more significance than James Thurber's far from untypical experience of failure to discern the structure of plant cells through a microscope. The strength of Galileo's case for the veracity of his telescopic observations of the moons of Jupiter derives from the range of practical, objective tests that his claims could survive. While his case fell short of being absolutely conclusive, it was incomparably stronger than any that could be made for the alternative, namely, that his sightings were illusions or artefacts brought about by the telescope.

4.6 Planetary sizes as viewed through the telescope

According to the Copernican theory the distance of a planet from
the Earth should vary appreciably during the course of the journeys
of each of them about the Sun. When a planet is on the same side of
the Sun as the Earth it will be relatively close, while when on the
opposite side of the Sun it will be relatively distant. In the case of
Mars the distance from Earth varies by a factor of the order of eight
and in the case of Venus by a factor of about six. Consequently, the
diameters of the planets as viewed from Earth should vary by a
similar factor. However, when viewed with the naked eye, Mars
appears to change size by a factor not much greater than two, while
the apparent change in size of Venus is negligible. It was for this
reason that Galileo (1967, p. 334) described Mars as launching a
'ferocious attack' on the Copernican system and Venus as presenting a 'greater difficulty'. When the two planets are observed
through a telescope the difficulty is removed. The observed changes
in size are in accordance with the predictions of the Copernican
theory.

Planetary sizes as viewed with the naked eye clash with the
Copernican theory whereas the corresponding telescopic data
conform with it. But which set of data is to be accepted? I will argue,
contrary to Feyerabend, that Galileo was able to make a strong case
for the telescopic data quite independently of the compatibility of
such data with the Copernican theory.

Galileo appealed to the phenomenon of irradiation to help
discredit naked-eye observations of the planets and as providing
grounds for preferring the telescopic observations. Galileo's hypothesis was that the eye 'introduces a hindrance of its own' when it
views small, bright, distant light sources. Because of this such
objects appear 'festooned with adventitious and alien rays'
(Galileo, 1967, p. 333). Thus, if stars 'are viewed by means of
unaided natural vision, they present themselves to us not as of their
simple (and, so to speak, their physical) size but as irradiated by a
certain fulgor and as fringed with sparkling rays' (Galileo, 1957,
p. 46). In the case of the planets, irradiation is removed by the
telescope.

Since Galileo's hypothesis involves the claim that irradiation
arises as a consequence of the brightness, smallness and distance of
the sighted source, it can be tested by modifying these three factors
in a variety of ways, many of which do not involve use of the

telescope. A number of ways are explicitly mentioned by Galileo. The brightness of stars and planets can be reduced by viewing them through a cloud, a black veil, coloured glass, a tube, a gap between the fingers or a pinhole in a card (Galileo, 1957, p. 46). In the case of planets, the irradiation is removed by these techniques, so that they 'show their globes perfectly round and definitely bounded', whereas, in the case of stars, the irradiation is never completely removed so that they are 'never seen to be bounded by a circular periphery, but have rather the aspect of blazes whose rays vibrate about them and scintillate a great deal' (Galileo, 1957, p. 47). As far as the dependence of irradiation on the apparent size of observed light sources is concerned, Galileo's hypothesis is borne out by the fact that the Moon and the Sun are not subject to irradiation (Galileo, 1967, p. 338). This aspect of Galileo's hypothesis, as well as the associated dependence of irradiation on the distance of the source, can be subject to a direct terrestrial test. A lighted torch can be viewed from near or far, and at day or night. When viewed at a distance at night, when it is bright compared with its surroundings, it appears larger than its true size. When viewed in the day, or close at hand, the apparent size is in conformity with the real size of the torch. Galileo appeals to this consideration to argue that his predecessors, including Tycho and Clavius, should have proceeded with more caution when estimating the size of stars.

> I will not believe that they thought that the true disc of a torch was as it appears in profound darkness, rather than as it is when perceived in lighted surroundings: for our lights seen from afar at night look large, but from near at hand their true flames are seen to be small and circumscribed (Galileo, 1967, p. 361).

The dependence of irradiation on the brightness of a source relative to its surroundings is further confirmed by the appearance of stars at twilight, which appear much smaller then than at night, and of Venus when observed in broad daylight which appears 'so small that it takes sharp eyesight to see it, though in the following night it appears like a great torch' (Galileo, 1967, p. 361).

This latter effect provides an approximate way of testing the compatibility of the Copernican (and other) theories with the observed sizes of Venus which does not involve an appeal to telescopic evidence. The test can be made with the naked eye provided observations are restricted to twilight. There are two reasons why this test will be difficult and not entirely satisfactory.

The first is that, under these conditions, Venus appears so small as to make accurate estimates of its apparent size difficult. The second is that it is not possible to carry out this test when Venus is close to its maximum and minimum apparent sizes because at those times it appears very close to the Sun. Consequently, it cannot be observed in the daytime because of the glare of the Sun, but only after the Sun sets, when Venus is close to the Earth and at its largest, or before it rises when it is furthest from the Earth and at its smallest. Nevertheless, according to Galileo at least, although the changes in size of Venus can only be precisely observed with the telescope, they are 'quite perceptible to the naked eye' (Drake, 1957, p. 131).

By fairly straightforward practical demonstration, then, Galileo was able to show that the naked eye yields inconsistent information when small light sources, bright compared to their surroundings, are viewed in the terrestrial or celestial domain. The phenomenon of irradiation, for which Galileo provided a range of evidence, as well as the more direct demonstration with the lamp, indicate that naked-eye observations of small, bright light sources are unreliable. One implication of this is that naked-eye observations of Venus in daylight are to be preferred to those made at night when Venus is bright compared to its surrounding. The former, unlike the latter, show that the apparent size of Venus varies during the course of the year. All this can be said without any reference to the telescope. When we now note that the telescope removes irradiation when used to observe planets and that, what is more, the variations in apparent size so revealed are compatible with the variations observable with the naked eye in daylight, a strong case for the telescopic data begins to emerge.

Our discussion of Galileo's method for measuring the motions of Jupiter's moons in Section 4.5 shows how Galileo was able to objectify and quantify his telescopic measurement of the apparent diameter of a planet during the year. The observed variations were precisely in accordance with the predictions of the Copernican theory. Galileo presented his telescopic observations of the apparent sizes of Mars and Venus as if they offered strong support for the Copernican theory. This was not justified. There was no question of the telescopic sightings of the apparent size of the planets offering support for the Copernican theory, as compared to its rivals, the Ptolemaic and Tychonian systems, because both the latter predicted precisely the same variations in size as predicted by Copernicus. Variations in distance from Earth, leading to predicted changes in apparent size, arise in the Ptolemaic system because the planets move closer then further from the Earth as they traverse the epicycles superimposed on the deferents, which latter do define paths equidistant from Earth. They occur in Tycho Brahe's system for the same reason as they occur in the Copernican system, since the two systems are geometrically equivalent. Derek J. de S. Price (1969) has shown quite generally that this must be so once the epicycles in the systems are adjusted so that they are compatible with the observed angular positions of the planets and Sun. That the apparent sizes of the planets had posed a problem for the major astronomical theories since antiquity is acknowledged by Osiander in his introduction to Copernicus's *Revolutions of the Heavenly Spheres*.

The telescopic observations of the changes in apparent size of the planets could not justifiably be used as evidence for the Copernican theory over its rivals, then. But these observations did provide a reason for accepting telescopic data in the astronomical domain in addition to those related to the phenomenon of irradiation. Unlike many naked-eye observations, the telescopic estimates of the sizes of the planets were consistent with all the astronomical theories of Galileo's time and acceptance of them removed a problem that had been present in astronomy since antiquity.

The foregoing discussion of Galileo's introduction of the telescope into astronomy enables us to put what has become known as the 'theory-dependence of observation' in perspective. It illustrates why a subjectivist reading of that thesis should be rejected. If we interpret 'objective' to mean something like 'testable by routine procedures', recognizing that the appropriate procedures will often require skills that few possess, Galileo was able to objectify his telescopic observations. What is more, they were able to withstand a range of tests, as we have seen. What is correct about the 'theory-dependence of observation' thesis is not that observation in science lacks objectivity, but that the adequacy and relevance of observation reports within science is subject to revision. Observation in science may be objectified, but we do not thereby have access to secure foundations for science. By the time Galileo's novel telescopic observations had become accepted by virtue of their ability to survive objective tests many previously acceptable observation reports based on naked-eye data became unacceptable because of their inability to survive tests made possible by Galileo's innovations.

One further example drawn from Galileo's science will reinforce my distinction between objective observation, which I believe to be achievable, and the availability of a secure, incorrigible empirical base for science, which I believe to be an empiricist myth. In his *Dialogue Concerning the Two Chief World Systems* Galileo (1967, pp. 361–3) described an 'objective' method for measuring the diameter of a star. He hung a cord between himself and the star under investigation and moved back and forth until the cord just blocked out the sun. Galileo argued that the angle subtended at the eye by the cord was then equal to the angle subtended at the eye by the star. We now know that Galileo's results were spurious. The apparent size of a star as perceived by us is due entirely to atmospheric and other noise effects and has no determinate relation to the star's physical size. Galileo's measurements of star size were theory-dependent and fallible and are now rejected. But this rejection has nothing to do with subjective aspects of perception. Galileo's observations were objective in the sense that they involved routine procedures which, if repeated today, would give much the same results as obtained by Galileo. In the next chapter, by reflecting on some characteristics of experimentation in science, I will reinforce this point that the absence of secure foundations for science is not due to problematic, subjective aspects of human perception.

Experiment

5.1 The production and rejection of experimental results

If there are to be secure foundations for modern scientific knowledge, as assumed by orthodox philosophers, then presumably it is
experiment, as opposed to mere observation, that provides them.
However, some general features of experimentation are such that
experimental results are quite unsuitable for constituting the secure
observational base sought by foundationalists. Experimental results are constantly rejected, revised, replaced or deemed irrelevant
for a variety of reasons which are quite straightforward from the
point of view of scientific practice. What is more, the extent to
which the experimental basis for science is constantly updated and
transformed has nothing much to do with problems associated with
human observation or perception. Even if the senses did provide us
with certain facts about the observable world, we would still lack
secure foundations for science. These points are obvious and
unproblematic once we adopt the standpoint of everyday scientific
practice as opposed to that of empiricist philosophy of science, as
the following examples will show.

My first example concerns the series of experiments performed
by Heinrich Hertz, in a two year period from 1886 to 1888, that
culminated in the first controlled production of radio waves (Hertz,
1962). Apart from revealing a new phenomenon to be explored and
developed experimentally, Hertz's results were of great theoretical

significance. They provided strong evidence for fundamental aspects of Maxwell's field theory of electromagnetism as opposed to the 'action at a distance' theories in vogue on the Continent. It was a consequence of Maxwell's theory that oscillating currents should radiate, although Maxwell himself did not appreciate this (Chalmers, 1973). In the main, Hertz's results and the significance he attributed to them remain acceptable from a modern point of view. However, some of his experimental data needed to be replaced and one of his main interpretations of them rejected. Both of these eventualities help to illustrate my anti-foundationalist case.

Hertz was able to use his experimental set-up to measure the velocity of the radio waves that he had produced. His results indicated that the waves of longer wavelength travelled at a greater speed in air than along wires, and faster than light, whereas Maxwell's theory predicted that they should travel at the speed of light both through air and along the wires of Hertz's apparatus. These results were inadequate for reasons Hertz already suspected. Waves reflected back onto the apparatus from the walls of the laboratory were causing unwanted interference. Hertz's (1962, p. 14) own comments on the problematic results were as follows:

> The reader may perhaps ask why I have not endeavoured to settle the doubtful point myself by repeating the experiments. I have indeed repeated the experiments, but have only found, as might be expected, that a simple repetition under the same conditions cannot remove the doubt, but rather increases it. A definite decision can only be arrived at by experiments carried out under more favourable conditions. More favourable conditions here mean larger rooms, and such were not at my disposal. I again emphasise the statement that care in making the observations cannot make up for want of space. If the long waves cannot develop, they clearly cannot be observed

Hertz's experimental results were inadequate because his experimental set-up was inappropriate for the task in hand. The wavelength of the waves investigated needed to be small compared to the dimensions of the laboratory if unwanted interference from reflected waves was to be removed. As it transpired, within a few years experiments were carried out 'under more favourable conditions' and yielded velocities in line with the theoretical predictions.

A point to be stressed here is that experimental results are required not only to be adequate, in the sense of being accurate recordings of experimental happenings, but also to be appropriate or significant. They must be designed to cast light on some significant question put to nature. Judgement about what is a significant question, and judgement about whether some specific experiment is an adequate way of answering it, will depend heavily on how the practical and theoretical situation is understood. It was the existence of competing theories of electromagnetism and the fact that one of the major contenders predicted radio waves travelling in air with the speed of light that made Hertz's attempt to measure the velocity of his waves highly significant, while it was an understanding of the reflection properties of the waves that led to the appreciation that Hertz's experimental set-up was inappropriate. These particular experimental results were rejected and soon replaced for reasons that are straightforward and non-mysterious from the point of view of physics.

As well as illustrating the point that experiments need to be appropriate or significant, and that experimental results are rejected or replaced when they cease to be so, this episode in Hertz's researches and his own reflections on it clearly brings out the respect in which the rejection of his velocity measurements has nothing whatsoever to do with problems of human perception. There is no reason whatsoever to doubt that Hertz carefully observed his apparatus, measuring distances, noting the presence or absence of sparks across the gaps in his detectors, and recording instrument readings. His results can be assumed to be objective in the sense that anyone who repeats them would get similar results. Hertz himself stressed this point. The problem with Hertz's experimental results stems neither from inadequacies in his observations nor from any lack of repeatability, but rather from the inadequacy of the experimental set-up. As Hertz pointed out, 'care in making the observations cannot make up for want of space'. Even if we concede to the empiricists that Hertz was able to establish secure facts by way of careful observation, we can see that this in itself was insufficient to yield experimental results adequate for the scientific task in question.

The above discussion can be construed as illustrating how the acceptability of experimental results is theory-dependent, and how judgements in this respect are subject to change as our scientific understanding develops. This is illustrated at a more general level by the way in which the significance of Hertz's production of radio

waves has changed since Hertz's time. At the time Hertz conducted them, Maxwell's theory, which understood electromagnetic phenomena as the manifestation of mechanical states of a
mechanical ether, predicted radio waves in a way that the rival
action-at-a-distance theories did not. Consequently, Hertz and his
contemporaries were able to construe the production of radio
waves as, among other things, *confirmation of the existence of the
electromagnetic ether*. Two decades or so later the theoretical
problem situation was importantly different. Maxwell's electromagnetic theory, suitably modified to incorporate the electron,
had ousted its action-at-a-distance rivals but was challenged by
Einstein's relativized version which dispensed with Maxwell's
mechanical ether. Both Einstein's and Maxwell's theory predicted
radio waves travelling at the speed of light. In this context, therefore, Hertz's production of those waves does not discriminate between the two theories, and so cannot be taken as evidence for the
existence of a mechanical ether. Hertz's experimental results are,
in the main, still accepted, but the significance attributed to them
has been transformed.

A second example, concerning nineteenth-century measurements of molecular weight, further illustrates the way in which the
relevance and interpretation of experimental results depends on
the theoretical context. Measurements of the molecular weights of
naturally occurring elements and compounds were considered to
be of fundamental importance by many chemists in the nineteenth
century, especially by those who favoured Prout's hypothesis that
the hydrogen atom is a fundamental building block from which
other elements are composed. The latter expected molecular
weights measured relative to hydrogen to be close to whole
numbers. The painstaking measurements of molecular weights by
the leading experimental chemists in the ninteenth century became
largely irrelevant from the point of view of theoretical chemistry
once it became realized that naturally occurring elements contain a
mixture of isotopes in proportions that had no theoretical significance. This situation inspired the chemist, F. Soddy, to comment
on its outcome as follows (Lakatos and Musgrave, 1974, p. 140):

> There is something surely akin to if not transcending tragedy in
> the fate that has overtaken the life work of this distinguished
> galaxy of nineteenth-century chemists, rightly revered by their
> contemporaries as representing the crown and perfection of

accurate scientific measurement. Their hard won results, for the moment at least, appear as of little interest and significance as the determination of the average weight of a collection of bottles, some of them full and some of them more or less empty.

Once again, we note that old experimental results become rejected as irrelevant, and for reasons that do not stem from problematic aspects of human observation. The nineteenth-century chemists involved were 'revered by their contemporaries as representing the crown and perfection of accurate scientific measurement' and we have no reason to doubt the adequacy of their observations and measurements. Nor need we doubt their objectivity. I have no doubt that if their experiments were repeated by those few contemporary chemists possessing the appropriate skills, similar results would be obtained. That they be adequately performed is a necessary but not sufficient condition for the acceptability of experimental results. They need also to be significant or relevant to some problem.

The points I have been making with the aid of examples can be summed up in a way that I believe is quite uncontentious from the point of view of physics and chemistry and their practice. The stock of experimental results regarded as appropriate testing ground for contemporary theory is consistently updated. Old experimental results are rejected as inadequate for a range of fairly straightforward reasons. They can be rejected because the experimentation involved inadequate precautions against possible sources of interference, because the measurements employed insensitive and outmoded methods of detection, because the experiments come to be appreciated as incapable of solving the problem at hand, or because the question they were designed to answer becomes discredited. While these observations can be seen as fairly obvious comments on everyday scientific practices, they nevertheless have serious implications for much orthodox philosophy of science, for they undermine the widely held notion that science rests on secure foundations. In so far as experimental results constitute the empirical evidence for our theories, they are constantly transformed and updated. Science does not have, and does not need, secure foundations. What is more, the reasons why it does not has nothing much to do with problematic features of human perception.

5.2 Implications for empiricism

One implication of my reflections on some common features of experiment in science has been stressed sufficiently in the foregoing section, namely, their incompatibility with the empiricist assumptions that secure foundations for science are provided by the senses. Observation statements, however securely established they might be considered to be by the senses, are alone inadequate for supplying experimental results significant for science.

As Roy Bhaskar (1978) has persuasively argued, experimentation is incompatible with a wide range of empiricist conceptions of scientific laws according to which they are construed as constant conjunctions of events, Humean fashion. According to those formulations, scientific laws accord with the schema 'whenever an event of type A occurs an event of type B follows' or, more in keeping with extreme empiricism, 'whenever an event of type A is observed to happen, an event of type B is observed to follow'. A problem for this view follows from my discussion, in Chapter 3, of the problem situation surrounding Galileo's introduction of experiment within physics. There are few, if any, observable regularities to be discerned in the observable world around us, so that, for example, strong contenders for such generalities as 'objects denser than water sink in water' are violated by floating needles and water insects. The natural world does not behave in a way that is sufficiently regular to allow exceptionless regularities to be discerned, although the solar system does come close to providing an exception. As our discussion of Galileo's innovations showed, there is a sense in which experimentation provides the answer to this problem. We can artificially construct physical situations in which regularities of the Humean type obtain, so that, for example, a particular change in current strength as displayed by an ammeter is always accompanied by the same displacement of a spot on a fluorescent screen. But if these regularities, which, in general, obtain only in artificial experimental situations, are identified with scientific laws, then we are at a loss to say what governs the behaviour of the world outside experimental situations. The constant conjunction view can be rendered compatible with the more orderly aspects of Hertz's experiments perhaps, but it does not permit an appeal to laws to explain how a radio signal of fluctuating strength reaches Sydney from the mid-Pacific. If scientific laws are identified with regularities, in the form of constant conjuctions, then irregular situations cannot be regarded as subject to laws. This clashes with the natural scientific assumption that irregular short-wave radio signals are governed by Maxwell's equations just as much as Hertz's radio waves were.

The above discussion highlights a problem for a particular empiricist conception of scientific laws. However, since Galileo at least, it does not pose a problem for science. Evidence bearing on the adequacy of scientific laws is acquired in artificial experimental situations, but the laws so identified are assumed to apply outside experimental situations also, although here their action will be superimposed on other laws, leading to irregular behaviour at the level of events. From the point of view of physics we have no problem understanding that surface tension intervenes to prevent a needle sinking in water or that various atmospheric and other disturbances lead to irregularities in the strength of a radio signal. Implicit in modern scientific practice is the assumption that natural phenomena are governed by laws, but that, in the natural world, those phenomena are juxtaposed in complex ways. This is why experimental intervention is necessary to unearth epistemologically relevant information. This is incompatible with the construal of laws as empirical regularities and also indicates why descriptions of observable states of affairs are in general quite inappropriate for constituting the building blocks from which scientific knowledge is constructed, as many empiricists would have it (cf. Feyerabend, 1981). Observable events are, in general, the outcome of a complex combination of diverse processes with no necessary epistemological significance. Science requires the production and observation of relevant events, and this is what experiment is intended to facilitate.

5.3   Implications for Popperian philosophy of science

A key component of Popper's elaboration of falsificationism is the notion of the empirical content of a theory. According to Popper, in science we seek theories with high empirical content, and a particular instance of theory change will be progressive if the newly accepted theory has a greater empirical content than its predecessor. The rationale underlying this construal of the aim of science is clear enough. If we think of the empirical content of a theory as a measure of the substantive claims it makes about the behaviour of the world, then a preference for theories with a high empirical content simply amounts to a preference for those theories that tell us a good deal about the world. Further, the more extensive the claims made by the theory, the more open it will be to possible falsification. Given two rival theories, opting for one with the greater empirical content is equivalent to opting for the more falsifiable one (Popper, 1972, pp. 112–13). When couched in these general terms, Popper's position appears plausible. However, when we look at the details of the way in which he elaborates it we find serious problems stemming from the role of experimentation as I have discussed it above.

Popper (1972, p. 120) defines the empirical content of a theory as the class of its potential falsifiers. A potential falsifier is a conjunction of observation statements (Popper calls them 'basic statements') which clashes with the theory. Thus, for instance, the conjunction of five observed planetary positions that do not lie on an ellipse would be a potential falsifier of the law that 'planets move in ellipses around the Sun'. Usually, potential falsifiers will involve the specification of an experimental set-up designed to test a theory together with the description of an outcome inconsistent with the prediction of the theory. A potential falsifier of Galileo's law of fall, for example, would be a description of the experimental set-up involved in his inclined plane experiment together with recordings of times of descent down various lengths of the plane inconsistent with a constant acceleration. By contrast, a description of the wayward descent of a falling leaf would not constitute a potential falsifier of Galileo's theory. The introduction of winds and air resistance renders the erratic fall compatible with Galileo's claims about unimpeded fall. The potential falsifiers of a theory will be those experimental results which are such that, should they occur, would refute the theory. The empirical content of a theory is identified with the set of events that it rules out. Scientific laws are prohibitions. Popper (1972, p. 113) explicitly asserts that theories tell us *nothing* about the events compatible with them.

Popper's identification of the content of a theory with the class of its potential falsifiers has an undesirable consequence. According to Popper (1972, p. 113), it is the class of its potential falsifiers that determines what a theory 'says' about the world and represents 'the empirical information conveyed by a theory'. However, as we have seen, except in exceptional circumstances such as those that obtain in the solar system, it is only by way of controlled experiment that a theory can be falsified, so that the class of potential falsifiers will be made up of the specification of experiments and their results. Popper's position entails that the content of a theory consists of the experimental results it forbids and so says nothing about the behaviour of the world outside experimental situations. This clashes with the fact that scientific theories are constantly applied outside experimental situations. The specification of the collapse of a bridge would not be a potential falsifier of Newtonian mechanics. The collapse would be attributed to fatigue, high winds and the like. Nevertheless, Newtonian mechanics is assumed to apply to the bridge by its designers, and with good reason. Similarly, the description of the uneven fall of a leaf in the autumn breeze would not constitute a potential falsifier of Newton's gravitational theory, but we nevertheless assume that gravity acts on the leaf in accordance with that theory during its fall, and we attribute the fact that autumn leaves usually end up on the ground to the action of gravity.

In identifying the content of a theory with the class of its potential falsifiers, Popper in effect identifies the domain of applicability of a theory with the domain of its adequate test situations. Elsewhere, Popper (1961, p. 117) expresses a more plausible view. The relevant passage reads as follows:

> The crucial point is this: although we may assume that any actual succession of phenomena proceeds according to the laws of nature, it is important to realize that practically no sequence of, say, three or more causally connected concrete events proceeds according to any single law of nature. If the wind shakes a tree and Newton's apple falls to the ground, nobody will deny that these events can be described in terms of causal laws. But there is no single law, such as that of gravity, nor even a single definite set of laws, to describe the actual or concrete succession of causally connected events; apart from gravity, we should have to consider the laws explaining wind pressure; the jerking movements of the branch; the tension in the apple's stalk; the bruise suffered by the apple on impact; all of which is succeeded by chemical processes resulting from the bruise, etc. The idea that any concrete sequence or succession of events (apart from such examples as the movement of a pendulum or a solar system) can be described or explained by any one law, or by any one definite set of laws, is simply mistaken.

Here Popper acknowledges that the falling apple is governed by causal laws such as the law of gravitation, but also acknowledges that the observed sequence of events cannot be described by any law, by any one 'definite set of laws'. The latter remark implies that descriptions of the sequence of events involved in the brief period of the apple's history do not constitute a potential falsifier of any causal law. Popper (1972) must therefore conclude that laws of nature do not 'say' anything or 'convey information' about the observed sequence of events. This is inconsistent with Popper's acknowledgement that 'nobody will deny that these events can be described in terms of causal laws'. My discussion of experiment suggests that this clash in Popper's writings should be removed by rejecting his account of the empirical content of a theory.

Popper hoped to explain the superiority of a theory over its predecessor in terms of the greater empirical content of the former. He appreciated that there were problems in making the notion of 'greater content' precise. He did offer some special cases for which the problem could be solved. Theory A can be said to have greater content than theory B if the potential falsifiers of B are a subclass of those of A. In this way we can capture, for example, the sense in which 'all planets move in ellipses' has a greater content than 'Mars moves in an ellipse'. However, once we appreciate, as I have urged above, that it is experimental results that constitute the testing ground for our theories, so that it is the specification of experimental arrangements and results that make up Popper's classes of potential falsifiers, then it can be seen that comparing rival theories by comparing their classes of potential falsifiers is inappropriate. For very often it will be the case that the experiments that provided the potential falsifiers of a succeeded theory, far from being incorporated into the class of potential falsifiers of its successful rival, will be dismissed as irrelevant for defendable reasons.

5.4 Defending experiment from sceptical attack

My account of experiment indicates that experimental results are theory-dependent in fundamental respects. Judgements about the adequacy and significance of an experimental result depend on high- and low-level theoretical assumptions concerning the appropriateness of the experimental set-up. Some draw highly sceptical conclusions from this. They conclude that experimental results cannot constitute an objective testing ground for our theories because they themselves involve theory. We are, as it were, trapped in the domain of theory and cannot get outside it to match our theories against data independent of them. Theories precipitate their own data, in the form of the experimental results they are used to sanction.

A very unambiguous expression of the above position is due to Barry Hindess (as quoted in Collier, 1979):

> If testing is a rational procedure then there must be an a-theoretical mode of observation governed by a pre-established harmony between language and the real. To maintain, as Popper does, both the rationality of testing and the thesis that observation is an interpretation in the light of a theory is to collapse into a manifest and absurd contradiction.

The implication is clear. If our observation reports and experimental results are theory-dependent then theory testing cannot be rational. A similar position is reached on a more general philosophical level in an influential work by Richard Rorty (1980). After a critique of foundationalist epistemology quite compatible with my own, Rorty concludes by advocating a highly relativist position according to which the search for knowledge is confined to the domain of 'conversation'.

The first necessary response to the sceptic here follows Ian Hacking's (1983) response to Rorty. While we can concede to the sceptic that all our observations and experimental reports and justifications of them are necessarily formulated in theory-dependent language, it is important to recognize that experiment involves, not simply talking about the world, but practically acting upon it. The second necessary response is to stress the strong sense in which the results of experiments are determined by the way the world is rather than by the theories that inform their design or interpretation or the experimenter's belief in those theories. While the details of an experimental set-up will depend on the theory-guided judgement of the experimenter, as will the significance attached to the results, once the apparatus is activated it is the nature of the world that determines the position of a pointer on a scale, the clicks of the geiger counter, the flashes on the screen, and so on. It was because the physical world is the way it is that an experiment conducted by Hertz in 1883 yielded no detectable evidence for the electromagnetic effect of cathode rays, just as it was because the world is the way it is that J. J. Thompson's more appropriate apparatus yielded detectable evidence two decades later

(Hon, 1987). It was the material differences in the experimental arrangements of the two physicists that led to the differing outcomes, not the differences in the theories held by them.

It is the fact that experimental outcomes are determined by the workings of the world rather than by the theoretical views of experimenters that provides the possibility of testing theories against the world. This is not to say that significant results are easily achieved, nor to deny that the significance of experimental results is sometimes ambiguous, nor to claim that experimental results and the conclusions drawn from them are infallible. I am arguing against sceptical relativism, not fallibilism. The aim to produce objective, unambiguous, significant experimental results represents a highly demanding challenge. While there is no a priori guarantee that the challenge can be met, the history of science and its practice indicates that it can often be met.

5.5   The experimenter's regress

In recent years, sceptical doubt has been cast on the role of experiment in science by sociologists. Recognizing the extent to which the adequacy of and significance to be attributed to experimental results relies on theoretical consideration and fine practical judgements, they conclude that there is a circularity involved when experiments are taken as providing an adequate testing ground for scientific theories. The problem has been referred to as 'the experimenter's regress'. Thus Andrew Pickering (1981, p. 229), in an analysis of experiments designed to detect quarks, writes:

> one cannot separate assessment of whether an experimental system is sufficiently closed from assessment of the phenomena it purports to observe: if one believes in free quarks then the Stanford experiment [which was deemed by its performers to have detected free quarks] is sufficiently closed; if not, then it is not.

Collins (1985, p. 84) makes a similar observation with references to experiments designed to detect high-flux gravity waves:

> What the correct outcome is depends on whether there are gravity waves hitting the earth in detectable fluxes. To find this out we must build a good gravity wave detector and have a

look. But we won't know if we have built a good detector until we have tried it and obtained the correct outcome! But we don't know what the correct outcome is until . . . and so on *ad infinitum*.

In the light of the existence of this circle, which he calls the 'experimenter's regress', Collins concludes that controversies in science cannot be settled by appeal to experiment in an objective, scientific way. 'Some "non-scientific" tactics *must* be employed because the resources of experiment alone are insufficient' (Collins, 1985, p. 143). Thus the demise of high-flux gravity waves was 'a social (and political) process' (Collins, 1981, p. 54). Not even experiments in the paranormal that are presented as revealing the emotional life of plants can be deemed non-scientific. If you believe in the paranormal, then the experiments are adequate, while if you do not they are not.

While I believe that the studies by Collins and like-minded sociologists throw interesting light on the nature and complexity of experimental work, I do not think their extreme conclusions are warranted. Nor are they supported by their own studies. To facilitate and clarify the discussion, I focus on one of the major studies that Collins employs to support his position, his investigation of the debate concerning experiments designed to detect gravity waves, from the time Joseph Weber claimed to have detected them in 1969 until the time the debate was ended and Weber's claims discredited by about 1975 (Collins, 1985, ch. 3).

The experiments were designed to identify signals due to the supposed interaction of gravitational waves with a detector, and to distinguish them from those expected from thermal and other noise. The strength of the signal that Weber claimed to have detected was such that it clashed by several orders of magnitude with what was to be expected according to theory accepted at the time, including Einstein's relativity theory. Weber's experiments were therefore treated with scepticism, especially as they operated near the limits of what could be regarded as statistically significant. The issue was not so much the existence of gravity waves, which were generally expected in the light of Einstein's theory, but the existence of the high-flux gravity waves that Weber claimed to have detected.

In the early 1970s attempts were made to replicate Weber's experiment, but these failed to detect statistically significant signals. There were two lines of investigation pursued by Weber that showed promise of strengthening his case. He claimed, firstly, that there were significant correlations between signals detected by detectors situated thousands of miles apart; and, secondly, a near twenty-four-hour periodicity in the signals which suggested a correlation between the signals detected and the orientation of the Earth with respect to the stars. Both of these correlations strengthened the claim that the signals detected by Weber were caused by a flux of gravitational waves striking the Earth from a particular direction in space. However, Weber's case for the correlation between separated detectors was seriously undermined by the discovery of an error in his computer program and the fact that some of the signals from distant detectors that he compared with his own, assuming them to have been recorded simultaneously, were in fact recorded four hours apart. As far as the sidereal correlation was concerned, Weber's attempts to substantiate it were unsuccessful. The correlation faded away.

A further factor in the debate between Weber and his critics concerned the type of system, including circuitry and computer program, used to process the raw signal from the detector. Weber's critics were able to draw on generally accepted knowledge to show that for a wide range of types of signal a linear system was more appropriate than the non-linear system employed by Weber. Weber could not achieve statistically significant results using a linear system. He concluded from this that the pulses he presumed to be caused by the absorption of gravity waves must indeed have an unusual profile. By about 1975 the scientific community became unconvinced by Weber's case, the existence of high-flux gravity waves was denied and that line of research abandoned.

Collins uses this study and others like it to challenge the distinctive epistemological status typically attributed to scientific knowledge. He concludes that complex debates within science cannot be settled by appeal to experiment by what are usually regarded as 'scientific' means. Rather, they are settled as a result of other social and political pressures. In the light of his case study of the dispute over gravity waves he concludes that there is 'no set of "scientific" criteria which can establish the validity of findings in this field. The experimenter's regress leads scientists to reach for other criteria of quality' (Collins, 1985, p. 88), so that 'some "non-scientific" tactics must be employed' (Collins, 1985, p. 143). He points out that there are possible ways of interpreting the case against Weber 'noting the flaws in each grain, such that outright rejection of the high flux claim was not the necessary inference' (Collins, 1985, p. 91). Because 'the experimenter's regress prevents an "objective" solution' (Collins, 1985, p. 151), it is the social and political interests of the scientific community that settle on one rather than another equally acceptable outcome. 'It is not the regularity of the world that imposes itself on our senses but the regularity of our institutionalised belief that imposes itself on the world' (Collins, 1985, p. 148).

Collins's claims are mistaken, and certainly not supported by his case studies. In particular, the experimenter's regress, as he and others such as Pickering construe it, is based on an inadequate understanding of the nature and role of experiment.

One point that needs to be stressed as a counter to extreme and unwarranted reactions to the theory-ladenness of experiment is the sense in which experimental outcomes are determined by the nature of the physical world rather than by the theories believed or entertained by experimenters or interpreters, as stressed in Section 5.4. Weber would have dearly loved the signals emerging from his experimental set-up to have exhibited a twenty-four-hour period-icity but the world did not co-operate.

We can accept, along with Collins and like-minded sociologists, that the adequacy and significance of an experimental result will be sensitive to background assumptions. The experimenter's regress, as formulated by Collins, and which threatens the assumption that experimental tests can offer an objective basis for theory appraisal, only has force if it is the case that the claims under test, for example, that there are high-flux gravitational waves or free quarks, form part of the background assumptions informing the experiments designed to test those very claims. If the adequacy of experiments designed to test for the existence of high-flux gravity waves can only be judged once a stand on their existence has been taken, then the kind of circularity invoked by Collins and Pickering does indeed result. However, this state of affairs is not the one typically faced by experimental scientists, nor does it correspond to the situation faced by Weber and his critics.

Given some dispute in science, the challenge is to arrive at definitive experimental results which do not prejudge the issue. Those experimental results will depend on background assumptions and these will be subject to challenge. If any of them are to be challenged in a way that is not idle or gratuitous, then the challenge should be accompanied by a strategy for discriminating between the challenged assumption and the proposed alternative. This is in keeping with the general aim of science as characterized in Chapter 3, according to which the adequacy of our claims about the world should be assessed by pitching them against the world in some practical way. The experiments carried out by Weber and his critics can readily be interpreted in this way. Judgements about the adequacy of the various results claimed involved a host of assumptions, but they were not of the kind that produces the circularity invoked by Collins. Some of Weber's results were discredited in a straightforward way by appealing, for example, to shared assumptions about what constitutes a reliable computer program. Other criticisms were more subtle. We noted above that Weber was criticized for amplifying his signals using a non-linear system when it was generally recognized that linear systems were more sensitive, given some weak assumptions concerning pulse shape. Weber accepted the background knowledge forming the basis of this criticism and concluded that the pulses emanating from his detector must have an unusual profile. His critics were correct to insist on the *ad hoc* character of this response. What would be required to strengthen Weber's case would be some independent evidence for the unusual profile. Such evidence could conceivably eventuate. More sensitive circuitry might enable the pulse shape to be diagnosed, for example. However, Weber and his supporters did not, and have not, produced such independent evidence. There are sound, objective scientific reasons for rejecting high-flux gravity waves on current evidence.

Most of the evidence Collins provides for the importance of 'non-scientific' factors in the controversy over high-flux gravity waves is derived from the results of interviews with the participants. Collins (1985, p. 87) shows that the reasons given by scientists for their acceptance and rejection of experimental results included such factors as the personality or nationality of the experimenters, the size and prestige of the university of origin, whether the scientists worked in industry or academia, the style of presentation of results, and so on. However, these observations provide no cause for concern even to the most orthodox construals of scientific rationality. The day-to-day decisions made by scientists concerning which lines of research to follow and the strategies to be adopted, which experiments to trust and which to question and so on, will, of course, be influenced by a host of subjective factors such as those pinpointed by Collins. But such factors should not, and, in the dispute over gravity waves, did not, determine the acceptability of scientific claims.

Another issue that Collins highlights using his inteview material is the variability and often contradictory nature of the beliefs and judgements of scientists. Thus one scientist regarded the fact that the statistical analysis in Weber's experiment was carried out by a computer to be a point in his favour, while another regarded it as a cause for concern; some regarded coincidences between separate detectors to be highly significant, others disagreed; some found the evidence for gravity waves convincing, others did not: Collins construes this evidence in a way that supports his view that there is no single correct scientific response to such issues, so that the fact that one response wins out over others must be explained by appeal to non-scientific factors. I accept, and am not surprised by, the variability in the judgements and beliefs of scientists noted by Collins. However, to transfer that variability to scientific knowl-edge itself is not justified, and is the result of too close an identification of scientific knowledge with the beliefs or opinions of scientists. Basically, what renders a scientific claim acceptable or utilizable is the extent to which it offers objective opportunities for future research or practical application, that is, the extent to which avenues for future investigation or exploitation present themselves, given the existing theoretical and technological resources (Chal-mers, 1982, ch. 11). The sociologist Karin Knorr-Cetina (1981, p. 8) makes a similar point when she writes:

> But where do we find the process of validation, to any significant degree, if not *in* the laboratory itself? . . . What *is* the process of acceptance if not one of selective incorporation of previous results into the ongoing process of research production? To call it a process of opinion formation seems to provoke a host of erroneous connotations. . . . What we have, then, is not a process of opinion formation, but one in which certain results are solidified through continual incorporation into ongoing research.

Weber and his supporters may have continued to believe very intensely in high-flux gravity waves and his opponents equally intensely in their non-existence after 1975, but this has little bearing on the fate of Weber's claim. The important point is that by 1975 there was little that could be done with Weber's claim. Once the attempts to substantiate the correlation between separated detectors and the sidereal correlation had failed, and once Weber had been forced to invoke untestable assumptions about pulse profiles, there was nowhere for him and his supporters to go, no objective opportunities to take advantage of, no way of solidifying his claim by incorporating it into ongoing research. This 'scientific' explanation of the lapse of interest in high-flux gravitational waves does not, and need not, invoke extra-scientific social or political interests.

Two qualifications need to be made with respect to this somewhat conservative response to Collins's challenge. Firstly, science is fallible, open-ended, and revisable. One can envisage ways in which Weber's claims might become resuscitated. Some advance in microelectronics might make possible the identification of the unusual pulse profiles postulated by Weber, opening up a range of opportunities for practical research. This in turn might open up opportunities for theoreticians to attempt to account for the waves detected or for astronomers to search for independent evidence of their source. Until and unless something of this kind happens, Weber's claims will remain in science's mortuary. Secondly, it must be acknowledged that there might well be episodes in science whose outcome is determined by social and political factors not operating in the interests of science, although I have argued, against Collins, that the episode involving Weber and high-flux gravity waves is not one of them. Many of the issues raised here, typical of those discussed by contemporary sociologists of science, are discussed more fully in the following chapters.

Collins's study of Weber's attempt to detect gravity waves illustrates the fact that the production of relevant experimental data in science is certainly not a straightforward matter. However, I have argued, against Collins, that the problems involved are not always insuperable and that objective experimental results having a critical bearing on our evaluation of scientific knowledge claims can and have been obtained. Hertz produced sound evidence for the existence of radio waves, while Blondlot failed to produce adequate evidence for the existence of N-rays and Weber failed to produce adequate evidence for the existence of high-flux gravity waves. These incidents, I maintain, can be adequately understood in terms of the aim of producing scientific knowledge and no resort to extra-scientific social or political factors is necessary for appreciating the epistemology of these situations. This is not to say that the aim of science can be practised in isolation from other aims and practices, nor that the aim of science does or should always prevail over other aims. Issues such as these are taken up in the remaining chapters of this book.

Science and the Sociology of Knowledge

6.1 Sociology and scepticism about science

A traditional view about the objectivity of science has it that the merits of a scientific theory are independent of the class, race, sex or other characteristics of the individuals or groups that espouse it. If influences stemming from such characteristics of individuals and groups are termed 'social' influences, then it can be said that, on the traditional view, the development and evaluation of science is not appropriately subject to a social explanation. Many contemporary sociologists deny that science is immune to social explanation in this sense, so that their views constitute a sceptical attack on the objectivity and distinctive epistemological status typically attributed to scientific knowledge. The following provide a few of the many possible examples of scepticism of this kind.

According to David Bloor (1982, p. 283), scientific laws are protected and rendered stable, not for reasons internal to science, but 'because of their assumed utility for purposes of justification, legitimation and social persuasion'; while David Turnbull (1984, p. 58) appeals to sociological studies to help defend his view that there is nothing distinctive about scientific knowledge and holds it to be 'subject to the same determinants and influences as other forms of knowledge'. The detailed study of laboratory work by B. Latour and S. Woolgar (1979, p. 237) led them to deny any useful distinction between science and politics whilst H. M. Collins and G. Cox (1976) explicitly defend an extreme relativist view of science according to which the strategies employed by Marian Keech to convince others of the reality of her communication with extra-terrestrial beings do not differ significantly from those employed in science.

A rebuttal of sceptical claims such as these will require a careful consideration of the senses in which science can be said to be subject to a social explanation. In this context a distinction is frequently invoked between what can be referred to as the 'cognitive' and 'non-cognitive' aspects of science. The latter involve such things as the social organization of science, the influence of science on other aspects of society, and the reverse influences which result in some branches of science being supported more than others. Larry Laudan, an opponent of current trends in the sociology of knowledge, invokes this distinction. He cites as examples of questions requiring a sociological answer, 'why a particular scientific society or institution was founded, why a scientist's reputation waned, why a particular laboratory was established when and where it was, or why the number of German scientists rose dramatically between 1820 and 1860' (Laudan, 1977, p. 197). Not even the most orthodox defenders of the autonomy and rationality of science would deny that there is a role for sociology in answering such questions. The existence of a legitimate non-cognitive sociology of science is not something that is disputed, although it must be said that its domain encompasses issues that are far less tame than those invoked by Laudan. If we include such problems as the impact of science on the environment, the potential of genetic engineering, the increasing gulf between technologically advanced and techno-logically undeveloped societies and the effect of computerization on our lives, then the non-cognitive sociology of science includes the most pressing social, political and moral problems of our time.

However its importance is evaluated, a legitimate domain for non-cognitive sociology of science is not in dispute. It is when we turn to the cognitive aspects of science that we come to the heart of the debate between traditional defenders of the autonomy and rationality of science and some contemporary sociologists. David Bloor opens his book *Knowledge and Social Imagery* (1976, p. 1) with the question 'Can the sociology of knowledge investigate and explain the very content and nature of scientific knowledge?' and proceeds to outline his 'strong programme in the sociology of knowledge' designed to give a positive answer to this question. He regards those sociologists who have stopped short of giving a social explanation of the content of science as suffering from a failure of nerve. Bloor, and a number of like-minded sociologists, have had the nerve to take the cognitive content of science as the object of their sociological explanations and their endeavours are typically interpreted by traditionalists as threatening the epistemological status of science (see also Mulkay, 1979, pp. 60–2; Mackenzie, 1981, pp. 2–4).

Yet another distinction is necessary before we can adequately pinpoint the terrain of the dispute. Roughly speaking, that distinction is between good science and bad science. Traditional opponents of the sociology of knowledge, while denying the appropriateness of a social explanation of the cognitive content of adequate science, are quite prepared to see external, social causes invoked when deviant or bad science is to be explained. Thus, traditionalists are more than ready to invoke social causes to explain the Lysenko affair in Russia or the corruption of physics in Nazi Germany but will not concede that it is appropriate to seek a social explanation of, for example, the replacement of classical by quantum mechanics. The readiness of traditionalists to accept a social explanation of inadequate science is evidenced by the extent to which they are at ease with explanations by anthropologists of the strange knowledge systems of alien tribes, such as the Azende belief in witchcraft, which appeal to aspects of the Azende social life.

Traditionalists and the more radical contemporary sociologists of knowledge are divided over the question of whether or not the cognitive content of our best science is subject to a social explanation. In the remainder of this chapter I attempt to take a discriminating look at the substance of the debate.

6.2 The sociologists' inadequate portrayal of their opponents

Attempts by sociologists to argue the need for a sociological account of the cognitive content of science, thereby undermining traditional views concerning the distinctive epistemological status of science, are frequently marred by an inadequate, outmoded, positivist-inspired portrayal of what those traditional views amount to. Mulkay (1979) paves the way for his version of the sociology of science by rejecting what he refers to as 'the standard view' of science, David Bloor (1976) presents his view as an alternative to some fairly extreme versions of rationalism and empiricism, and Barry Barnes (1977) develops his position in opposition to the 'contemplative account' which involves an extreme version of the correspondence theory of truth based on an analogy with pictures. I am certainly content to join the sociologists in the rejection of these views. But so, for example, does Karl Popper, whom the above-mentioned scholars would hardly welcome as a fellow sociologist of knowledge. There are far more sophisticated attempts to defend the distinctive status of science than those that the sociologists typically take to constitute the opposition.

An example which illustrates my point is the way in which Mulkay (1979) sets up his programme by criticizing the 'standard' account. Elements of the standard view identified by Mulkay are as follows. Science is able to establish truths about the natural world in the form of universal laws of nature. These laws are confirmed by appeal to factual statements established by careful, unprejudiced observation. While some theoretical components of science might go beyond what can be established by observation, a clear distinction can be drawn between the theoretical and observational levels. At the latter level science exhibits cumulative growth. The criteria by which knowledge claims are to be judged are universal and ahistorical. The conclusions of science are determined by the physical world rather than by the social world.

Mulkay devotes the second chapter of his book to the rejection of this standard view. He appeals to an argument of Hanson's (1969) to urge that the physical world cannot be demonstrated to be governed by universal laws and that the arguments typically offered to establish this are circular. He rehearses the various ways in which the traditional distinction between observation and theory is inadequate and illustrates the revisability of empirical evidence. He insists that criteria for judging the merits of theories are not universal but context-dependent and subject to change. In so far as the criteria are social products, the conclusions of science are not simply determined by the nature of the physical world.

Mulkay is surely right to reject what he calls the 'standard' view, but he misdescribes it as standard, since only a minority of contemporary philosophers of science wishing to defend the epistemological status of science would disagree with him. For instance, most of Mulkay's rejections are not only compatible with Popper's philosophy of science, they form the most distinctive features of it. The fact that Popper denies that scientific theories can be proved and insists on their permanent hypothetical character hardly needs documenting. Further, he rejects the notion of a secure basis for science and insists that observation statements are theory-laden and revisable (Popper, 1972, ch. 5). He stresses that observation and experiment are properly conceived of as active interventions in rather than passive receptions of nature (Popper, 1979, Appendix 1). He stresses the importance of context-dependent decisions in the acceptance or rejection of observations and experimental results (Popper, 1972, pp. 104–6). He also observes that knowledge is a social product produced by the modification of earlier knowledge and not established directly through a confrontation with the physical world (Popper, 1979, p. 71). Popper could perhaps be said to adhere to the 'standard' view of science in so far as he subscribes to a correspondence theory of truth, while if we take his falsifiability criterion as specifying an absolute demarcation between science and non- or pseudosciences he might be interpreted as denying the context-dependence of some scientific standards. However, we only have to turn to Imre Lakatos, another philosopher antipathetic to the sociology of knowledge and who saw himself as developing Popper's ideas, to find one who dispenses with a correspondence theory of truth (Hacking, 1983, ch. 8) and who explored ways in which scientific standards have historically changed (Lakatos, 1978b). Popper and Lakatos are typical of a host of contemporary philosophers who reject Mulkay's standard view and who attempt to give a more adequate defence of the distinctive epistemological status of science. Consequently, a challenge of that epistemological status will require much more than a rebuttal of discredited traditional views.

Another argument that figures prominently in the writings of sociologists of science, and which betrays their inadequate characterization of the opposition, runs as follows (see, for example, Barnes and Bloor, 1982, p. 23; Bloor, 1982). Scientific theories are underdetermined by the evidence. *Therefore* extra-scientific social factors enter into the processes that lead to the selection of one among the perhaps many possible theories compatible with the evidence. A particularly clear example of this line of argument is to be found in an interesting paper in which David Bloor (1982) attempts to reinstate and apply to science the thesis of Durkheim and Mauss that 'the classification of things reproduces the classification of man'. Bloor employs Mary Hesse's network model to illustrate the complex way in which scientific statements relate to each other and to the world. Hesse uses the term 'correspondence conditions' to refer to the ways in which scientific statements are constrained by the empirical evidence and the term 'coherence conditions' to refer to other constraints. Bloor (1982, p. 283) urges that it is in the coherence conditions that we are to locate the entry of social relations into science. He utilizes themes found in the work of the anthropologist Mary Douglas to argue that 'certain laws are protected and rendered stable because of their assumed utility for purposes of justification, legitimation and control'.

This move, from the underdetermination of theories by the evidence, to the presence of interests other than knowledge producing interests in science, is much too swift, and concedes much too much to the traditional accounts of science to which the sociologists are opposed. The logical point that there is an infinity of universal statements compatible with a given, finite, set of observation statements leads traditional empiricist philosophers of science to the conclusion that there is an infinity of scientific theories compatible with given evidence. This, of course, is totally at odds with situations encountered in actual science, where scientists often struggle to find *any* workable theory compatible with some problematic evidence. The underdetermination argument pays insufficient attention to the growth of science. New knowledge arises as a response to problems arising with previous knowledge. If novel theories are to be intelligible there is no option but to use existing concepts or to modify or extend them by analogies with other existing concepts, while if they are to be of any use they must offer the promise of some possible feasible line of investigation. Attempts to analyse such notions in terms of simplicity (Popper, 1972, ch. 7), coherence and progressiveness (Lakatos, 1974) or degree of fertility (Chalmers, 1982, ch. 11) are indicative of the fact that underdetermination cannot be presumed to lead necessarily to the introduction of extra-scientific social factors into science.

The position that I have defended in earlier chapters of this book involves a view of science that renders its practice social in fundamental respects. Observation reports and experimental results are human, social products that emerge as the result of argumentation and experimentation. However, in general, their acceptance, and, where necessary, their rejection or transformation, can be understood in terms of the aim of science without resorting to broader social factors. In the previous chapter I attempted to show that this was the case in the controversy surrounding attempts to detect gravity waves, thereby countering Collins's more radical sociological construal of that episode. While scientific results are not 'determined by the physical world' as a result of some face-to-face confrontation, as the extreme empiricist would have it, experiments are designed to enable the physical world to play a crucial role in the acceptance or rejection of those results. I have argued, especially in Chapter 2, that the methods and standards of science are historically contingent, social, products subject to change, but I attempt to offset the extreme relativist consequences that might be thought to follow from this by indicating how such changes can be understood in terms of the aim of science, a point I illustrated with my account of Galileo's introduction of the telescope into astronomy. If sociologists of science are to argue for the social determination of the cognitive content of science in a way that offers some grounds for scepticism about the objectivity and distinctive epistemological status typically attributed to it they must do more than combat extreme and largely outmoded philosophies of science.

6.3   The social origins of scientific knowledge

If we are to evaluate claims to the effect that the content and nature of scientific knowledge are subject to sociological explanation then we will need to be clear about just what it is that is to be explained and what an explanation amounts to. One way of construing the claim is to understand the explanation of some instance of scientific knowledge as involving the historical story of how that knowledge was constructed. If we understand the claims of the sociologists in this way, then I am prepared to concede that the content of scientific knowledge is subject to a sociological explanation. It is frequently the case that concepts and practices employed to good effect in science have their origins in the social world outside scientific practice narrowly conceived. A sociological account of the origins of scientific knowledge is often appropriate.

The path that led Darwin to his theory of evolution provides a good example. Darwin's view of natural selection was heavily influenced by Malthus's idea that the size of human populations has a natural limit, because unlimited growth will lead them to outstrip the food supply. His proposal was a contribution to social debates of the time connected, among other things, with the problem of poverty. Darwin's arguments for the transmutation of species and for the mode of that transformation drew on and were influenced by acquaintance with the techniques of professional breeders. There is no doubt that an adequate account of the emergence of the theory of evolution up until Darwin's mature theory and beyond, takes us far beyond the confines of scientific discourse to include broader social factors (Young, 1969; 1971).

Turning to the physical sciences, a second example is provided by the kinetic theory of gases, introduced by James Clerk Maxwell in the nineteenth century. The statistical techniques that Maxwell employed to deduce the macroscopic properties of gases from the random motions of the constituent molecules drew on techniques devised by social theorists to deal with regularities in social phenomena such as the birth or crime rate (Porter, 1981).

If to explain a component of scientific knowledge is to give a full and adequate account of how it emerged, then we can readily concede that a range of factors typically dealt with by sociologists will be relevant, so it can be said that there is a legitimate role for a sociology of scientific knowledge here. However, there is another kind of 'explanation' of scientific knowledge that can be sought. We can seek to explain, and evaluate, how, and to what extent, an instance of scientific knowledge functions as such. We can consider the extent to which it contributes to the aim of science. Thus, referring back to our Darwinian example, we can seek to identify the account of selection and evolution present in Darwin's writings. We can raise questions about its internal consistency and its relationships to evidence and we can compare it with rivals in this respect. Such questions are not only legitimate, they are precisely the ones that are significant if we are interested in the epistemological status of Darwin's theory. What is more, the answers to them are independent of considerations about the social origins of Darwin's ideas. As it happens, Darwin's own theory is not beyond criticism from an epistemological point of view. In particular, Darwin's own writings do not make adequately clear just what the mechanism of selection is, and how that postulated mechanism is born out by evidence. This point is especially important because, in Darwin's time, the fact that evolution occurs and has occurred was generally accepted. What was in dispute was the correct account of the mechanism of evolution (Young, 1971).

In Chapter 3 I attempted a cautious articulation of the aim of modern science. I claim that it not only makes sense to interpret and evaluate Darwin's theory from that point of view, but further, that that aim was as a matter of fact adopted and striven for within the biological practice of the time. It was the aim of evolutionary theorists of the time to produce an adequate account of the mechanism of evolution, although they also participated in other practices, such as religion and politics, with different aims. Questions concerning the adequacy of Darwin's theory as an account of the mechanism of evolution are distinct from questions about its origins or about the various ideological uses to which the theory was put. If sociologists of knowledge wish to claim that an explanation of how a theory functions as knowledge and how it contributes to the aim of science must involve social factors other than those internal to science, then I take issue with them.

The position I adopt here can be construed as a version of the traditional distinction between the so-called mode of discovery and the mode of justification. According to that distinction, the way in which a theory comes to be proposed is one kind of question, requiring a historical answer, while the way in which it is to be justified as adequate knowledge is another kind of question, requiring an epistemological answer. I have no objection to my position being characterized in this way provided a number of qualifications are recognized. Firstly, the mode of justification is delimited by me in terms of an account of the aim of science rather than by reference to a specific definition of scientific method or rationality. Secondly, some historical questions are relevant to the mode of justification, as Lakatos and his followers have stressed (Musgrave, 1974b; and Nickles, 1987). The need for a theory to constitute an advance on the theory it challenges, and the importance of novel predictions in this context, introduces a historical element into the domain of justification. Thirdly, my claim that the aim of science, and corresponding epistemological questions, can be *distinguished* from other aims and other kinds of question should not be taken as implying that the activity of producing scientific knowledge can be *separated* from other activities, an issue that I will take up in Chapter 8. Fourthly, the distinction between questions of origin and questions of scientific worth should not be taken as devaluing investigations of the former. The way in which scientific innovation can occur, and the way in which advances in a specialized science can come about through an input from outside the speciality, have important implications, for example, for the institutional organization of science and for science education.

6.4 The inappropriate emphasis on belief

Frequently, the debate between sociologists of scientific knowledge and their opponents takes place on the assumption that what is to be explained is the beliefs of scientists. Laudan (1981, p. 173), for instance, takes the sociologists he opposes to be claiming 'that we can give a *sociological* account of why scientists adopt virtually *all* of the beliefs about the world which they do'.

I have indicated elsewhere (Chalmers, 1982, chs 10 and 11) why I side with Popper in regarding a focus on the beliefs of individuals as quite inappropriate for understanding the nature of science and its progress. We are rarely in a position to know much about the degree of belief a scientist has in the theory on which he or she works, nor do we need to know if we are concerned to characterize and evaluate the scientific character of that work. I have no idea to what extent Weber's belief in high-flux gravity waves was affected by the research described in the previous chapter. My characterization and evaluation of that episode stand or fall on considerations of claims made, arguments offered and experiments performed rather than on considerations concerning the beliefs of the scientists involved. It is not uncommon for scientists to work on theories which they do not believe in an attempt to discredit them, and they sometimes make a contribution to the development of those theories in the process. An example is the support Poisson inadvertently gave to Fresnel's wave theory of light in the nineteenth century. Poisson's attempt to discredit the theory by demonstrating that it had the 'absurd' consequence that a bright spot should be observed at the centre of the shadow side of a suitably illuminated opaque disc backfired when that bright spot was observed experimentally. Given some of the problematic features of contemporary quantum mechanics, I am not sure what a belief in it would amount to, but I am reasonably clear about what it means to develop it, to compare it to classical mechanics in various respects and to test its consequences experimentally.

The inappropriateness of a focus on the beliefs of scientists when attempting to characterize science has been given strong support by a contemporary sociologist of science. Karin Knorr-Cetina (1981,

p. 8), on the basis of her studies of laboratory work, insists that it is quite inappropriate to consider the development of science in terms of the formation of the beliefs of scientists. According to her, a scientific result becomes accepted, not as a result of scientists choosing to believe it, but as a result of being incorporated into the 'ongoing process of research production', so that 'to call it a process of opinion formation seems to provoke a host of erroneous connotations'.

As long as we continue to identify scientific knowledge with the beliefs of scientists then we are inevitably forced into a version of the traditional debate concerning the extent to which beliefs are to be attributed to reasons or causes. According to the traditional standpoint, beliefs are rational to the extent that they are formed in the light of good reasons, and irrational to the extent that they are brought about by psychological and sociological causes. Laudan (1977, p. 198) subscribes to a version of this distinction in the course of his critique of the sociology of science:

> The intellectual historian of knowledge will generally seek to explain why some agent believed some theory by talking about the arguments and the evidence for and against the theory and its competitors. The cognitive sociologist of knowledge, on the other hand, will generally try to explain why the agent believed the theory in terms of the social, economic, psychological and institutional circumstances in which the agent found himself. Both are trying to solve the same problem (namely, the belief of some historical agent), yet their modes of solution are so different as to be almost incommensurable.

Laudan's view is that the cognitive content of adequate science should be explained by appeal to reasons, and that sociological causes need only be invoked when science goes astray. Social, or 'external' history of science is subservient to intellectual, 'internal' history of science (Laudan, 1977, p. 208).

By identifying a scientific theory with the belief of some historical agent, it seems to me that Laudan has chosen a very unsuitable terrain on which to defend his view. As I indicated above, we are rarely in a position to appreciate what the beliefs of scientists actually are, and, whatever the case may be in this regard, I am sure that those beliefs and their intensity will be influenced by a wide range of psychological and sociological factors as well as by arguments and reasons. Even the belief of an agent in one of

Laudan's main examples of a rational belief, $2 + 2 = 4$, will be influenced by the way he or she is taught it and by the derision that any attempt to deny it invokes. I find Laudan's suggestion that William Charleton may have accepted the mechanical philosophy for purely rational reasons most implausible.

There is plenty of scope for a sociological study of the beliefs of scientists and their connection with such things as class background. However, bearing in mind the distinction between scientific knowledge and individual belief, such studies do not in themselves constitute a sociological account of the cognitive content of science. There remains the problem of the relation between the beliefs of scientists and the cognitive content of the scientific knowledge they produce and develop. Once again, my position is supported by Knorr-Cetina when she writes (Knorr-Cetina, 1983, p. 116):

> even if we learn convincingly what particular individuals or groups believe in a set of propositions, we have not received an answer to the question whether and how these propositions in themselves embody social factors, nor to the question whether and how social factors influence the survival of and acceptance of knowledge claims. In other words, the epistemological question how that which we come to call knowledge is constituted and accepted is not addressed . . .

The discussion so far in this chapter has led us to acknowledge that there is scope for a sociological analysis of the origins of scientific knowledge, of the beliefs of scientists and of 'non-cognitive' aspects of science. Such analyses can be important and far from trivial. However, they stop short of offering a sociological explanation of the cognitive content of science in the sense of demonstrating how instances of scientific knowledge function as knowledge. It remains to assess the traditional position that allows a sociological explanation of bad, but not of good, science.

6.5 Sociological explanation restricted to bad science

It is commonly claimed that a sociological account of the cognitive content of science is appropriate only in those cases where science has gone astray. Accordingly, when science proceeds successfully its progress is explicable in terms of its own internal, 'rational' dynamic, so that a sociological explanation appealing to external influences is both unnecessary and inappropriate. Laudan and Lakatos have both recently supported versions of this claim. According to the former, 'the sociology of knowledge may step in to explain beliefs if and only if those beliefs cannot be explained in terms of their rational merits' so that 'the application of cognitive sociology to historical cases must await the prior results of the application of the methods of intellectual history to those cases' (Laudan, 1977, pp. 202 and 208). According to the latter, 'the *rational* aspect of scientific growth is fully accounted for by one's logic of scientific discovery' which may have to be supplemented by external explanation only to explain 'the residual, non-rational factors' (Lakatos, 1978a, p. 118).

David Bloor is just one of a number of contemporary sociologists who vehemently take issue with what they see as this unjustified attempt to limit the scope of sociological explanation. Bloor (1976, pp. 6–7) characterizes the stance that he opposes using propositions such as 'nothing makes people do things that are correct but something does make, or cause them to go wrong' so that 'the rational aspects of science are held to be self-moving and self-explanatory. Empirical or sociological explanations are confined to the irrational'. I find Bloor's discussion unhelpful, because of the extent to which it gives an extreme, uncharitable and largely unjustified rendering of the position of his opponents. In the following I offer a defence of a version of the traditional view according to which certain kinds of sociological explanation of the cognitive content of science are inappropriate. However, my own position certainly does not conform to Bloor's caricatures, nor is it identical to the positions defended by Laudan and Lakatos.

The following analogy will help to illustrate my position. Suppose that a game of soccer is under way and imagine that the ball lands at the feet of a player in front of the unguarded goal of the opposing side. In this context we would not regard the action of the player in kicking the ball into the goal as something requiring an explanation, or, rather, we would regard an 'internal' explanation as self-evident given the rules of soccer. On the other hand, if the player, rather than dispatching the ball into the goal, produced a knife and fork and attempted to eat it, this would make no sense in the context of the game. Some external explanation would be called for, perhaps invoking information about the mental health of the player. This is an extreme example, of course, but it does illustrate the way in which a legitimate distinction can be drawn between internal and external explanation. In a context in which actors engage in a practice with specific aims, then when their activity contributes to those aims no explanation that looks beyond the nature of the practice itself is called for. This is not to say that the game of soccer is some God-given activity not subject to any kind of explanation. A range of questions about the origins of the game, the psychological and social functions it serves, the economics of its professionalization, and so on, can be legitimately raised. There are certainly contexts in which a sociological explanation of soccer is called for. Nevertheless, in a context where the game and its rules are taken for granted, the actions of the players are appropriately understood internally unless they cannot be reconciled with the aim of the game.

A similar kind of point can be made with respect to the extreme positions targeted by Bloor in his attempt to defend a symmetrical approach in his sociology of knowledge. Some views on perception expressed by W. Hamlyn possess the asymmetry abhorred by Bloor. According to Hamlyn, 'the ways in which we may perceive something can be divided into two classes – the right ways and the wrong ways. Indeed, one way of perceiving something – the right way – may be distinguished from all the others'. The right way 'provides no room for scientific explanation, since none is called for'. If two lines of equal length are seen as such 'nothing *makes* them look of equal length' because 'they just *are* so' (Bloor, 1981, p. 205). I can agree with Bloor's rejection of these claims of Hamlyn, in so far as they deny that human perception is susceptible to any explanation. It is perfectly legitimate to ask how it is that human perception functions as it does, both when it is successful and when it deceives us. However, it is not difficult to modify Hamlyn's position in a way that preserves an asymmetry, while avoiding the claim that adequate perception is, in some way, its own explanation. In a context where the mechanism of perception is taken for granted, no special explanation need be invoked to explain why people see what they see. In such a context, if Macbeth claims he sees a dagger before him, no explanation is called for when there is a dagger present, while an 'external' explanation is called for, perhaps invoking Macbeth's psychological state, if there is no dagger present. There is certainly an asymmetry here, although it is not adequately characterized by Hamlyn.

While analogies between science, on the one hand, and soccer or human perception on the other, have their limits, they do serve to illustrate the way in which adequate science is to be understood and explained internally, by reference to the aim and point of the activity. Questions such as why the wave theory of light supplanted the particle theory, why Blondlot's claims about N-rays and Weber's claims about high-flux gravity waves were rejected by the scientific community, and how and why the results of Hertz's electrical researches were so rapidly incorporated into the practice of physics, are questions most appropriately answered internally, by reference to the aim of science, the aim to produce general knowledge able to come to grips with the nature of the world in a superior and more extensive way than previous knowledge. To seek an external answer to them in terms of the class background, nationality or other characteristics of the scientists involved is as inappropriate as seeking a similar type of explanation of why a soccer player takes effective advantage of an open goal. There is a strong sense in which the traditionalists are right to insist that the merits of a theory should be assessed independently of the psychology, class background or other characteristics of its proponents.

The claim I make for a legitimate domain for internal history of science and for internal, non-sociological, explanation and appraisal does not compel me to deny any explanation for science nor to regard science as its own explanation, proceeding according to an eternal, God-given mode of rationality. The existence and extent of the practice of science in our society and its interrelations with other social, political and economic practices are matters requiring analysis and explanation. As I have mentioned, and as I will elaborate in Chapter 8, these issues, far from being trivial, involve some of the most pressing political and social problems of our time. As for the methods and standards implicit in the practice of science, these are subject to change, and any such change requires explanation. However, in a context where the aim of science is adopted such changes can be explained internally, by reference to practical and theoretical discoveries and developments, rather than externally, by reference to class interests and the like. Of course, if any position that allows for a change in method or standards and denies an eternal, universal mode of rationality is deemed a sociological position, then I must count myself as a sociologist of science. In such a circumstance, what distinguishes me from the more radical sociologists is the extent to which I insist that science, its methods and its mode of progress, can and should be understood internally in terms of its general aim to produce knowledge rather than in terms of other aims or interests. This is not to adopt the naïve view that science can be practised in isolation from other interests, nor that those other interests never, or even should never, impede the realization of the aim of science. I simply insist that it is possible and important to distinguish the aim of producing scientific knowledge from other aims, and that the distinction is essential for an adequate explanation and appraisal of science.

In the next chapter I attempt to make the foregoing, somewhat abstract, considerations more concrete by taking a critical look at two detailed studies in which attempts are made to explain the cognitive content of science sociologically.

— 7 ———————————————

Two Sociological Case Studies

———————————————

7.1 Statistical theory and social interests

The first case study I analyse is Donald Mackenzie's (1978 and 1981) investigation of the influence of social interests on the development of statistical theory in late nineteenth- and early twentieth-century Britain, a study that is frequently cited as an exemplary one of its kind (Barnes and Mackenzie, 1979; Shapin, 1982). Mackenzie attempts to defend a very strong version of the sociology of knowledge. As he points out (Mackenzie, 1981, p. 2) 'no one doubts that there must be *some* relationship between science and the social context in which it develops'. He proceeds to distinguish between a weak and a strong version of that relationship. According to the former, social influences can influence such things as the pace of scientific advance and the direction in which social support is channelled. As far as influencing the content of science is concerned, according to the weak version of the sociology of science, social influences can only distort science by diverting it from its proper path. To the extent that social influences penetrate into the content of science, bad science results. According to the strong version of the sociology of science, social influences can affect the content of good science. Mackenzie endeavours to exemplify the strong version by showing how social interests affected the content of mathematical statistics in Britain around the turn of the century.

The social interests invoked by Mackenzie in his sociological explanation are those of the professional middle class at the time in question. Although Mackenzie does not claim to use the notion of class in a technical Marxist or any other sense, the nature and membership of the professional middle class is reasonably clear. It consists of those who work for a wage, rather than those who live off capital, but who are distinguished from the proletariat in that their work involves skilled mental, rather than manual, activity. Entry into the class is by way of education and training rather than birthright through inherited wealth or aristocratic status. The professionals were the custodians of areas of knowledge and expertise and their power derived from the extent to which that knowledge and expertise played an important social role. It was in the interests of the professional middle class to maximize the importance of that role, while maintaining strict control over its membership.

Eugenics, as it was developed in turn-of-the-century Britain, could be, and was, used to serve the interests of the professional class. According to that social theory, 'civic worth', identified with 'mental ability', was a fixed, natural characteristic of each individual human that was inherited. Only those who possessed a high degree of this natural characteristic were able to survive the demands of a professional training. The professional class could thus be seen as *naturally* superior, not only to the working class, who could be perceived as naturally and appropriately such because of the lack of mental ability of its members, but also of the aristocratic class and business communities, since the acquisition of wealth or inheritance of aristocratic lineage was no guarantee of mental ability. A social hierarchy, with the most able professionals at the top, could be construed as a natural hierarchy from the viewpoint of eugenics.

The substantive claims of eugenics concerning hereditary and social worth were typically augmented by a social programme designed to alter the genetic composition of the human race for the better. For instance, various measures were proposed to discourage or prevent breeding among the extremely poor, criminals and the mentally defective while tax incentives and family allowances were proposed to encourage a high birth rate within the professional class. The eugenics programme served to enhance the power of the professionals who possessed the knowledge of what was seen as the natural processes underlying social ones.

Let us accept that, subject to qualifications to which Mackenzie (1981, pp. 46–50) himself draws attention, eugenics provided an opportunity for the professional middle class to enhance its interests. The next step in Mackenzie's argument involves the connection between eugenics and developments in mathematical statistics. Articulation and documentation of the mode of inheritance assumed in eugenics required the development of appropriate statistical techniques. It is by analysing such developments in the hands of proponents of eugenics such as Francis Galton and Karl Pearson that Mackenzie aims to offer a strong case for the social determination of science. His study is meant to bear witness to the entry of the interests of the professional middle class into the very content of mathematical statistics. Let us examine the extent to which he succeeds, concentrating on the work of Galton and Pearson.

Francis Galton lived his life among Britain's intellectual elite. He himself reports that his early thoughts on heredity were influenced by the kinship links he noted among the intellectuals at Cambridge. He convinced himself that the kinship links among those of exceptional mental ability were more extensive than those to be expected if mental ability were randomly distributed. Galton's early thoughts on heredity certainly seem to have had their origins in some features of his social experience. The theoretical context in which Galton developed his theoretical ideas on heredity was that of naturalism, a standpoint that had flourished after Darwin and which constituted an area in which professional scientists struggled to win a domain of influence from religious authority. Galton explicitly wrote in terms of replacing religious authority with a 'scientific priesthood' (Mackenzie, 1981, p. 55).

Galton was able to draw on existing error theory for statistical techniques necessary for his eugenic concerns. Errors in a measurement were understood statistically to fluctuate about a mean value following what we would now call a normal distribution. Galton adapted these techniques to deal with the variability of human characteristics, such as height, among members of a population. However, rather than merely adapting error theory, Galton needed to extend it, and here he made fundamental contributions to mathematical statistics. For the quantitative account of descent that he sought, Galton needed to be able to deal with the statistics of dependent variables. In particular, he needed to cope with relationships between the distribution of a variable (such as height) in successive generations. It was in this context that Galton developed the concepts now referred to as regression and correlation in bivariate normal distributions.

Galton's eugenics, and his statistics, were taken up and developed by Pearson. The latter was very much part of the professional middle class of intellectuals. He espoused a brand of socialism similar to that of the Fabians, aiming for reforms that would see power based on the wealth of the bourgeoisie replaced by power based on knowledge and mental skills. Eugenics fitted well into this programme, as we have seen, and Pearson came to see eugenics and socialism as inseparable. As Professor of Applied Mathematics at University College, London, Pearson collaborated with W. F. R. Weldon, Professor of Zoology, to attempt to put Darwinian evolutionary theory on a firm mathematical foundation. A biometric laboratory and a eugenic laboratory were set up and the 'house journal', *Biometrika*, was initiated. Pearson eventually became Galton Professor of Eugenics, a chair funded by money from Galton's estate earmarked for that purpose.

Pearson made major contributions to mathematical statistics, refining Galton's techniques and extending them to multivariate distributions. Mackenzie (1981, ch. 7; 1978) attempts to illustrate the extent to which Pearson's eugenic concerns, and the social interests they served, entered into the heart of his technical statistical work by analysing a dispute between Pearson and one of his former pupils, Gill Yule. The dispute concerned the correct way to measure associations between data related to the biological world, especially human characteristics. For variables, such as height, that were continuous, measurable and normally distributed, correlation coefficients could be constructed in what was by then a straightforward and uncontentious way. The problem concerned data related to phenomena that were not measurable on a continuous scale, such as eye colour and intelligence. Pearson developed measures for association between such data on the assumption that underlying them were some variable factors distributed in a normal or some other regular way. Yule regarded such an assumption as unwarranted. Indeed, for the range of discrete variables in which he was particularly interested (e.g., alive or dead, inoculated or not inoculated) he regarded Pearson's assumption as absurd. He devised pragmatic measures of association for two such variables arranged in two by two tables (e.g., vaccinated or not, alive or not) that suited his practical needs. Pearson regarded Yule's measures as theoretically insignificant, and pointed out that the actual measures of degree of association varied, depending on which of a number of differing measures employed by Yule were adopted. Yule responded that if he struck to the same measure during the course of one particular investigation his measures met the practical needs they were designed for and led to no inconsistencies. Mackenzie explains this divergence of views in terms of the interests at stake. Pearson's commitment to his measures is attributed to the kinds of correlations involved in his eugenic assumptions, while Yule's are attributed to his more pragmatic interests concerned with ameliorating social problems among the poor. Mackenzie does not attempt to explain Yule's stand in terms of broader social interests, but does suggest that a concern with removing causes of unrest among the poor matched the interests of the downwardly mobile class to which Yule's family belonged so that we have 'the possibility that specific social interests sustained the non-eugenic statistics of Yule and his supporters' (Mackenzie, 1981, p. 182).

The above illustrates the kind of case to be found in Mackenzie's writings, although, of course, I have omitted much interesting detail. I suggest that Mackenzie's account of the entry of social interests into scientific practice exemplifies a weak version of sociological explanation rather than the strong one to which he aims to give substance. In particular, Mackenzie does not show social interests entering the content of mathematical statistics in a way that is sufficiently strong to support his case (cf. Yearley, 1982; Woolgar, 1981).

While it is true that the contributions to statistics made by Galton and Pearson came about in the context of investigations of heredity with implications for eugenics, those advances had a quite general application. Galton himself carried out statistical investigations of the weight of sweetpea seeds and of the stature of humans, for example, neither of which had direct relevance to eugenics. With respect to some of Pearson's innovations, Mackenzie (1981, p. 90) notes that his definitions 'were indeed general, but it is clear that man was the organism to which they were primarily intended to apply'. This implies that the social interests were present in Pearson's intentions, rather than in the statistics itself. Mackenzie acknowledges that many came to work with Pearson to learn skills they could apply in areas remote from eugenics. W. S. Gosset, for instance, applied methods of partial and multiple correlation developed in Pearson's school to improve brewing techniques, thereby enhancing the fortunes of Arthur Guinness and Son for whom he worked (Mackenzie, 1981, pp. 111–13). The fact that the mathematical statistics useful for eugenists, and hence serving the interests of the professional middle class, could also be used to serve the interests of the bourgeoisie, is not compatible with the claim that the statistics embodied professional class interests in a strong sense.

If we turn from mathematical statistics to the eugenics it was used to develop, then it is possible to identify the presence of social interests in the content of the latter. Many of the assumptions central to eugenics, even when seen in the narrower context of a theory of hereditary, rather than in the broader one of a social programme, had little justification when appraised from the point of view of the production of knowledge. The view that human beings possess an inherent characteristic 'civic worth' and that this characteristic was distributed in a normal or some other regular way was simply assumed rather than argued for. Evidence invoked to support the eugenic assumptions, such as the observation that, in the main, children of the intellectual elite tended to become members of that elite in their turn, could readily be subject to a social, environmental explanation. However, little or no research was carried out to discriminate between these competing explanations. I have no doubt that much of the content of eugenics is to be explained by reference to the social interests it served, *as opposed to* the extent to which it functioned as knowledge. But this constitutes a social explanation of 'bad science', in keeping with weak, as opposed to strong, sociological explanation.

Even if I am correct to deny that the content of mathematical statistics is not appropriately explained by appeal to broad social interests, there is much about that practice that warrants a sociological explanation, and Mackenzie makes valuable contributions towards that end. It is surely correct to say that an explanation of why advances in statistics were made when they were, and the extent to which that activity gained social support and an institutional footing, are strongly connected with eugenics and the extent to which it served the interests of the professional middle class of the time. The precise form that sociological explanations of such matters are to take is not easy to specify and I do not think that Mackenzie succeeds in clarifying this issue. He explicitly rejects the idea that his sociological explanations are intended to explain the psychology or motivations of individuals and he rejects a determinist view according to which an individual's ideas are caused by his or her social background (Mackenzie, 1981, p. 92). Here

Mackenzie quite rightly takes issue with the characterization of sociological explanation utilized by Laudan in his critique of the sociology of knowledge. According to Laudan (1977, p. 217),

> any cognitive sociological explanation must, at the least, assert a causal relationship between some belief, x, of a thinker, y, and y's social situation z. It will (if the explanations of sociology are in any sense 'scientific') do so by invoking a general law which asserts that all (or most) believers in situation type z adopt beliefs of type x.

Not only are explanations conforming to this pattern unavailable in sociology, they are, in general, unavailable in any other science. (If an autumn leaf falls to the ground we can invoke gravity to explain why. But not all leaves fall to the ground. Many of the leaves from the trees in my garden are carried up onto the roof and clog my gutters). What is more, the discussion of the previous chapter indicates why I regard Laudan's focus on individual belief as inappropriate, a point with which Mackenzie sometimes, but by no means consistently, seems to agree.

Mackenzie fails to supply an adequate, general characterization of the form of his sociological explanation. He tells us that his social analysis points to 'a match' of beliefs and social interests (Mackenzie, 1981, p. 92). He also asserts that 'we can . . . sometimes usefully discuss individual beliefs in social perspectives' (Mackenzie, 1981, p. 73). However, such remarks can be interpreted in a very weak sense, and are hardly adequate characterizations of a strong programme in the sociology of knowledge. Elsewhere, Mackenzie's analysis is more in keeping with his non-individualist assertions, involving an analysis of the institutionalization of science.

I suggest that Mackenzie's analysis of the social interests at work in connection with the development of statistics in Britain between 1865 and 1930 is best accommodated in something like the following way. First, our social analysis should attempt to understand the social situation in such a way that various groups or classes and their interests are identified. Here I have no particular quarrel with the way in which Mackenzie identifies the professional middle class and its interests. Having done this, it is possible to identify ways in which eugenics provided opportunities that could be exploited in the interest of that class. Once it is also recognized that the development of eugenics required developments in mathematical statistics, we are in a position to understand how developments in the latter provided opportunities for furthering the interests of the professional middle class. This, I suggest, is as far as a general analysis can go. Within this framework claims such as 'statistics flourished in Britain at the turn of the century because it provided opportunities for serving the interests of the professional middle class' have explanatory force while not requiring that the beliefs and motives of particular individuals be identified and deduced from their social situation.

The extent to which various opportunities were taken advantage of, by whom and in what way, is a contingent matter that can only be filled in as a result of historical research, and is not subject to a general sociological explanation. Mackenzie's analysis fills in such contingencies in a number of ways. For example, his story of how Pearson was born into an upwardly mobile middle class, how he reacted to the poverty and squalor of Victorian England and to the 'complacent superficiality' (Mackenzie, 1981, p. 75) of Cambridge University, how he became acquainted with various brands of socialism during a visit to Germany, and so on, shows how Pearson came to be in a good position to further his class interests by developing mathematical statistics. Translating this into terms employed by Mackenzie, we can appreciate that the 'match' between eugenics and the interests of the professional middle class provided an opportunity for furthering those interests which Pearson, given the nature of his mathematical training, was in a good position to take advantage of and, as a matter of fact, did take advantage of.

In summary, then, I concede to Mackenzie that there is room for a social analysis of the practice of mathematical statistics in Britain during the period in question and that he makes useful contributions to such an analysis, although there is room for clarification concerning the precise form his social explanations take. This is sufficient to offset a purist, conservative view that the pursuit of knowledge in academic institutions proceeds according to its own dynamic with no connection with broader political or social interests. The material support for the development of statistics at University College, London, was intimately bound up with the eugenics movement, as Mackenzie shows. What is more, the eugenic theories, as opposed to the mathematical statistics, served the interests of the professional middle class to a much greater degree than they served the aim to produce knowledge. However, whatever importance is attributed to Mackenzie's analysis, I deny that he has offered a social explanation of the content of mathematical statistics that is sufficient to establish his strong claim about the sociological determination of good science.

7.2   Freudenthal's social explanation of Newton's *Principia*

As we have seen, a weakness of Mackenzie's attempt to explain mathematical statistics socially is that he fails to make clear just what form his social explanation of cognitive content is meant to take. He fails to provide an adequate answer to Knorr-Cetina's question of how the theoretical propositions themselves embody social factors (Knorr-Cetina, 1983, p. 116). The same cannot be said of Gideon Freudenthal's (1986) construction of a social explanation of some aspects of Newton's physics. Freudenthal is not content to point to parallels or matches between scientific theories, on the one hand, and social relations or social conceptions, on the other. Rather, he endeavours to trace the precise way in which social relations enter into the content of Newton's physics. Let us investigate the extent to which he is successful.

Freudenthal does not seek to give a social derivation of the entire content of the *Principia*. He does not seek to explain Newton's laws of motion and his law of gravitation socially. Nevertheless, Freudenthal does seek to demonstrate how other, significant, assumptions at work in the *Principia* have their origins in, and are sustained by, social relations. The path Freudenthal traces from social relations to the cognitive content of Newton's science is roughly as follows. The social change from feudalism to early forms of capitalism engenders a conception of society according to which the latter is to be understood in terms of the essential properties of the individuals of which it is composed. This form of explanation becomes transformed into a general philosophical principle according to which the properties of wholes are to be explained in terms of the essential properties of their parts. This principle, when applied in the context of Newtonian physics, has determinate effects on some of its content. Following Freudenthal, I consider some of the detail of this process of entry of social relations into physics in the opposite order to that in which it is alleged to have occurred, beginning with the identification of those aspects of the *Principia* that are to be explained socially.

First among Freudenthal's targets for a social explanation is Newton's conception of an absolute space, for which Newton offered arguments in the *Principia* – the famous bucket experiment and the related one involving the rotation of two particles connected by a spring. The deformation of the water surface in the rotating bucket and the extension of the spring joining the particles were taken by Newton to indicate the presence of rotation relative to an absolute space existing independently of matter. Next among Freudenthal's targets we have Newton's distinction between essential and universal properties of matter. Universal properties are possessed by all material bodies encountered in empirical or experimental situations, as are essential properties. But some stronger requirement is apparently necessary for a property to count as essential, in Newton's view. Newton makes it explicitly clear that, for example, extension is a universal and also an essential property of bodies although gravity, while being a universal property, is not an essential property. Thirdly, Newton defined 'quantity of matter' as the product of density and volume, whereas elsewhere in the *Principia* density is defined as mass per unit volume. There is an apparent circularity here if we make the natural identification of 'mass' and 'quantity of matter'. The fourth target for Freudenthal's social explanation that I consider is Newton's argument from the fact that materials differ in density to the conclusion that they must contain vacuous spaces in varying degrees.

All of the contentions of the *Principia* noted in the foregoing paragraph are problematic. The experiments with a rotating bucket or particle pair could be interpreted as indicating motion relative to the stars, say, while even if we do follow Newton and assume that absolute motion is established, this is still insufficient to establish Newton's conclusion that the motion takes place in a space that is independent of matter. With respect to the distinction between universal and essential properties, it is difficult to see what empirical consequences could possibly follow from the claim that a property is essential, over and above being present in all bodies observed or experimented on. The apparent circularity involved in Newton's discussion of density is clearly a problem, while there are a number of ready alternative explanations for differing densities to the one Newton takes to be necessary.

Freudenthal argues that there is an assumption presupposed in the *Principia* which, if granted, removes the above difficulties. That assumption is that the material world is composed of equal particles each possessing the same essential properties, properties that a particle would continue to possess even if it were alone in empty space. I shall, for convenience, refer to this as the 'elementary-particle assumption'. It certainly helps Newton's argument for absolute rotation if it is presupposed that it is meaningful to conceive of the bucket rotating in an otherwise empty and pre-existing space. In the light of the elementary-particle assumption we can understand why gravity is denied the status of an essential property by Newton even though it is present in every body encountered in the world. A particle alone in empty space, while it would still possess extension, say, would not, according to Newton, possess gravity. If we understand 'quantity of matter' as 'number of elementary particles' then quantity of matter is, indeed, volume multiplied by the density of particles. But since the particles cannot be directly observed and counted, quantity of matter, and hence density, in this sense cannot be measured. However, since we can measure, or compare, masses and volumes, we can give an operational definition of density as mass divided by volume. There is no circularity because there are two conceptions of density involved, only one of which is measurable. Finally, once it is assumed that differing materials are made up of equal elementary particles, then differing densities do require the existence of varying degrees of space between the particles, as Newton concluded.

A strong form of the elementary-particle assumption is not stated explicitly by Newton in the *Principia*, although there is indirect evidence for Newton's espousal of it there and elsewhere in his writings. A weaker version explicitly formulated by Newton reads as follows (Freudenthal, 1986, p. 22):

> The extension, hardness, impenetrability, mobility, and force of inertia of the whole, result from the extension, hardness, impenetrability, mobility and forces of inertia of the parts; and hence we conclude the least particles of all bodies to be also all extended, hard and impenetrable, and moveable, and endowed with their proper forces of inertia. And this is the foundation of all philosophy.[1]

We note, with Freudenthal, that Newton does not argue for this assertion. He states it as if were evident. As far as the stronger claim, made explicit by Freudenthal in the form of the elementary-particle assumption, is concerned, the main case for the fact that Newton did indeed assume it is that, once assumed, otherwise problematic arguments and assertions in the *Principia* make sense, as we have seen. A plausible explanation of why Newton never did make explicit all the components of the elementary-particle assumption, and why no arguments are to be found for them in his writings, is that he regarded them as evident. That is, he took them for granted.

If we accept Freudenthal's reconstruction of the elementary particle assumption and its role and status, then we are in a position to appreciate the target of Freudenthal's social explanation of the *Principia*. The assumption that the material world is to be explained in terms of the essential properties that make it up, where an essential property is understood as a property that a particle would possess were it alone in empty space, functions as an evident principle in the *Principia* and, although it had no effect on the testable empirical content of physics at the time, the principle had determinate effects on the substantive claims made. How is this state of affairs to be explained? How did the elementary-particle assumption come to be taken as evident? These are the questions Freudenthal sets out to answer by reference to social relations.

Freudenthal traces Newton's assumption back to the individualist conceptions of society that emerged in the seventeenth century as feudal society gave way to early forms of capitalist society, with the market coming to play an increasingly fundamental role. We start from the fact that feudal society became increasingly unviable with the growth of towns and the increased interdependence of towns and of countries. The increased importance of the market that followed the increase in complexity and interdependence made it possible for merchants to acquire wealth and power, not by birthright, but by taking advantage of opportunities provided by the market, while it became increasingly possible for peasants to leave the land and the jurisdiction of the lord of the manor to become wage labourers in the towns. The emerging capitalist societies needed to be understood and justified. An alternative to the feudal conception of naturally ordered hierarchical society, given its classical formulation by Thomas Aquinas, became a theoretical and political necessity. Thomas Hobbes responded to the challenge early in the seventeenth century, especially in *Leviathan*, and alternatives were soon to follow later in the century, notably the one formulated by Newton's contemporary, John Locke. Freudenthal draws attention to the by now well-appreciated fact that, while the various conceptions of society formulated by the various theorists differed in fundamental respects, they shared one feature in common. All of them attempted to explain society by reference to the essential properties of the individuals that make it up, properties that individuals were assumed to possess independently of their existence in society.

The analysis thus far indicates a striking parallel between the relationship between individual and system as they figured in Newton's physics, on the one hand, and conceptions of society that emerged and became accepted in the seventeenth century, on the other. However, Freudenthal makes it clear that he is not content with the drawing of parallels as if they amounted to an explanation. Rather, having noted the emergence of an individualist conception of society as a response to social changes, he wishes to plot the precise path along which a version of that individualism entered Newton's physics. That path, according to Freudenthal, was by way of philosophy. Freudenthal attempts to show how Newton followed Hobbes in extracting from social theory a general philosophical conception of the relationship between element and system which he came to regard as evident and subsequently applied in his physics.

Freudenthal stresses the extent to which Hobbes's theorizing should be seen as a political programme designed to combat feudal social relations and the hierarchical conception of society that was used to justify it, and to facilitate the emergence of the new form of society. We must reject the view that Hobbes's ideas were an unconscious reflection of the contract relations evident in the marketplace since, in Hobbes's time, feudal relations persisted and were experienced by Hobbes, albeit as something to be opposed and replaced. He seized on contractual relations as they existed between independent proprietors in the marketplace and asserted them to be *the* basis for an analysis of society. He embarked on a project to substantiate his theoretical claims, a project that had political implications for the steps to be taken to replace feudal social relations with relations based on contracts between free and independent individuals.

In the seventeenth and eighteenth centuries it was common to distinguish three branches of philosophy: social philosophy; natural philosophy; and *philosophia prima* (metaphysics or first philosophy). The last of these was regarded as a body of abstract generalizations applicable to both social and natural philosophy. Freudenthal notes that Hobbes's assumption that society should be understood in terms of the essential properties of the individuals of which it is composed became a general philosophical principle in Hobbes's writing. That is, the assumption became part of first philosophy. It was applied by Hobbes to the physical world where, for instance, he considered the properties a body would have if created anew in empty space, concluding that it would have the property of extension only. The generalization of the assumption concerning the relation between element and system to first philosophy, and thence to natural philosophy, also helped Hobbes's political programme in so far as it served to undermine the alternative view of the relation between element and system permeating medieval philosophy, social theory and natural science, in which, of course, systems were theoretically prior to their elements. Hobbes's theorizing constituted an attack on the notion of hierarchy central to feudal social relations on three fronts: first philosophy, social philosophy and natural philosophy.

Hobbes's political programme was successful in so far as the relations between element and system which he introduced into his social theory, and thence into first philosophy and natural philosophy, became generally accepted. Freudenthal documents this with reference not only to Newton but also to others such as Jean-Jacques Rousseau and Adam Smith. Perhaps we can sum up Freudenthal's social explanation of the acceptance of the principle that wholes are to be understood in terms of the essential properties of their parts as follows. That principle was accepted because it served the interests of those who adopted and promulgated it and because it could be readily and convincingly exemplified by appeal to the character of exchange relations within the increasingly important marketplace as well as by appeal to mechanical analogies, such as the explanation of the properties of a watch in terms of the properties of its parts.

In so far as Freudenthal's story is adequate, we have now reached the stage where we can understand how Newton came to adopt and regard as evident the notion that a system is to be understood in terms of the essential properties of its parts. Freudenthal takes the analysis further to indicate how some of the details were worked out in Newton's first philosophy and how, at that level, a link was forged between his science and the individualist conception of society. In his philosophy Newton argued that our experience of voluntarily moving a limb establishes as self-evident the fact that we, as humans, possess freedom of will and that the matter that we choose to move is itself passive. Newton's argument purports to establish at one stroke both the conception of freedom as an essential property of individuals and passivity as an essential property of matter.

Other details of the social analysis are traced to the specific social situation confronted by Newton. The society of independent proprietors envisaged by Hobbes was not to eventuate. Rather, a capitalist society emerged, with most land and other means of production owned by a few. This did not occur without a political struggle involving the suppression of the Levellers, a struggle in which Newton himself played a partisan role. The culmination of that struggle in England involved a compromise with the King. The limited power that was to remain with the King was justified on the grounds that it was necessary for the maintenance of a social order that would otherwise not eventuate. In effect, then, the system, in this case society, cannot be totally accounted for by appeal to the essential properties of its parts. Outside intervention is necessary. We find precisely the same kind of situation portrayed in Newton's physics. The physical properties of the world system cannot be attributed to the physical properties of the corpuscles that make it up. Because of the inelasticity of collisions between corpuscles, and because of the motion introduced into the world through our voluntary actions, the total amount of motion will not be conserved automatically. Nor can gravity be explained by reference to the essential properties of bodies, as we have noted. In both cases we find Newton and his supporters appealing to divine intervention. God is governor of the world clock just as the King is governor of society.

I can accept the general line of Freudenthal's account of how individualist theories of society arose as a response to social change and how these became transformed, by many, into a general philosophical principle that wholes are to be explained in terms of the properties of their parts. However, as is clear from Freudenthal's account, that much is as applicable to Hobbes as it is to Newton. Since Hobbes's physics was significantly different from Newton's (for example, Hobbes regarded his particles as possessing the essential property of extension only) something needs to be added to mere individualism or atomism for Freudenthal to be able to complete his social explanation. As we have seen, Freudenthal added some details of Newton's philosophical conception of free will and his view of God as governor of the world, both sets of views being traced to aspects of Newton's political stance on social issues of his day. These views of Newton's were indeed widely accepted and exploited by those low church Anglicans and Whig politicians who occupied similar social positions or who adopted similar political stances as Newton's (Jacob, 1976). But they were not universally adopted. I suggest they can best be regarded as ideological extensions of Newton's physics rather than as parts of it. This point is strengthened by the fact that other physicists were able to construe Newton's physics in ways that differed radically from those aspects of Newton's own interpretation explained socially by Freudenthal. For example, Clerk Maxwell departed radically from the fundamental particle assumption when he utilized Newtonian mechanics to develop his electromagnetic field theory, in which localized phenomena are understood in terms of the mechanics of a continuous, all-pervasive material medium whilst Thomson and Tait (1879, p. 222) departed from Newton's conception of matter as passive, understanding it to have 'an innate power of resisting external influence'.

Freudenthal's analysis cannot be taken as a social explanation of the cognitive content of successful science. Nor, incidentally, does Freudenthal present it as such. He explicitly distinguishes between those aspects of the *Principia* that have a scientific justification (such as, for example, his laws of motion) and those assumptions that do not. It is precisely the latter that he attempts to explain socially. It might be objected here that I am using the benefit of hindsight to distinguish the good and bad parts of Newton's physics. Let us imagine, for the sake of argument, that the assumptions socially explained by Freudenthal were eventually vindicated. Let us imagine that contemporary science makes possible the measurement of motions relative to absolute space and the detection and counting of Newton's corpuscles. I would suggest that an appropriate response would be to say that, while certain assumptions had their origins in seventeenth-century social changes and social theory, they only received an adequate scientific interpretation and justification centuries later. We would have a situation similar to the one which traces some of Darwin's innovations to the writings of Malthus and the social situation that inspired them.

Freudenthal's analysis of Newton's *Principia*, which I chose as an example because it was the best, most careful and detailed social explanation of science that I could find, cannot be taken as a successful social explanation of the cognitive content of good science. Nevertheless, this by no means trivializes the importance and interest of his study and others like it. In so far as it is successful, the analysis shows how easily assumptions having social and political origins and serving social and political interests can enter into science under the guise of adequate science. Not even Newton's *Principia*, which might be expected to provide an excellent example of pure science, was free of such incursions. It can by no means be taken for granted that all that is put forward in the name of science and justified ostensibly in terms of its interests and aims does, as a matter of fact, serve those interests and contribute to that aim, and this is as true today as it was in Newton's time.

7.3 Concluding remarks

At this stage it is appropriate for me to make some general remarks to summarize the outcome of my critique of the sociology of science in this and the preceding chapter.

The natural world does not behave in one way for capitalists and in another way for socialists, in one way for males and another for females, in one way for Western cultures and another for Eastern cultures. A large-scale nuclear war, made possible by modern science, would destroy us all, whatever our class, sex or culture. But does it not follow from such platitudes that, in so far as science involves the attempt to construct generalizations that adequately characterize the natural world, the adequacy of such characterizations is quite independent of the predispositions or interests of the individuals or groups that construct and espouse them?

The radical sociologists who wish to defend a sceptical view of science might respond to the above observations as follows. The conception of generalizations about the world whose adequacy is assessed independently of sociological characteristics of the individuals or communities that construct and defend them is, at best an unrealizable ideal and at worst a nonsensical one. Knowledge claims and evidence that are produced, and the criteria by which they are assessed, are social products, and, as such, are inevitably shaped by social interests. Because of the kinds of social being that we are and the modes of constructing and testing knowledge available to us, interests, such as class interests, inevitably enter into our science.

Put in this way, the claims of the sociologists can be interpreted as empirical claims. As such, I regard them as false. I claim that the scientific community has been able to develop methods and techniques for constructing and testing knowledge claims that can, and often have, contributed objectively to the aim of science. My discussion of observation and experiment in Chapters 4 and 5 was designed to indicate how objective tests of the adequacy of knowledge claims can be devised in practice, and I showed this to be the case in situations chosen by the sceptics themselves as favouring their position. I showed, for instance, how the transformations of facts and standards accomplished by Galileo, used by Feyerabend to support a radical scepticism, can be understood as an objective step forward from the point of view of the aim of science; I also argued that the rejection of Weber's experimental claims with respect to gravity waves, used by Collins to illustrate the way in which external social and political interests enter into science, can be understood in terms of their failure to survive objective tests and legitimate criticism. Weber was left with nowhere to go. Objective progress towards meeting the aim of science can, and has been, made, which is not to say that it will prove possible in all instances, nor that it is achieved by unchanging methods or with respect to unchanging standards.

Sociological case studies, such as the ones described in this chapter, indicate how interests other than those serving the aim of science can influence the practice of science. There are no grounds for complacently assuming that scientific practice in fact proceeds in a way that is determined solely, or even mainly, by the aim to produce adequate knowledge. The practice of science is inevitably interconnected with other practices which have other aims and serve other interests. However, I suggest that an adequate understanding of this situation is not helped by ignoring or contesting what I believe to be a fairly straightfoward distinction between the aim to produce adequate scientific knowledge and other aims.

7.4 Note

1 I take issue with Freudenthal on one point of detail. Those of Newton's arguments that Freudenthal attempts to render cogent by introducing the

elementary particle assumption require only that the particles be of the same density and not, as Freudenthal insists, that they should also be of the same size. Further the use of the plural 'forces of inertia' by Newton in connection with the particles suggests that he did not assume they were all of the same size. Correcting Freudenthal on this point of detail does not effect the main thrust of his argument, however, which is why I restrict my criticism to a footnote.

— 8 ——————————————————

The Social and Political
Dimension of Science

————————————————————————

8.1   Introductory remarks

The gist of my stance with respect to extreme relativist or sceptical
construals of science can be summed up as follows. The aim of the
natural sciences is to extend and improve our general knowledge of
the workings of the natural world. The adequacy of our attempts in
this regard can be gauged by pitching our knowledge claims against
the world by way of the most demanding observational and
experimental tests available. While there is no universal method or
set of standards to preside over this quest for knowledge, and while
there is always present the possibility that the aim will be thwarted
by the surreptitious entry of other interests with different aims, the
aim can be, and often is, attained. The natural world is the way it is
independently of the class, race or sex of those who attempt to know
it, and the scientific merit of the theories that constitute our
attempts to characterize it should be similarly independent of those
factors. In spite of the social character of all scientific practice,
methods and strategies for constructing objective, albeit fallible and
improvable, knowledge of the natural world have been developed
in practice and have met with success.

Apart from the denial of universal method and the acknowledgement that science is fallible and its practice inherently social, the
above remarks can be read in a decidedly conservative way. It might
be concluded in the light of them that I do not regard a political and
social analysis of scientific practice to be appropriate in any strong

sense, and it might be surmised that I believe that all is well in contemporary science and that this will remain so as long as it remains autonomous and is protected from political and social influences. This is far from my position, and I devote this final chapter to an attempt to put the matter straight.

8.2   Objective opportunities and individual choice

In a nutshell, my main point is this. While the aim of science can be *distinguished* from other aims and epistemological appraisals distinguished from other appraisals, the scientific practice involved in the pursuit of that aim cannot be *separated* from other practices serving other aims. I move towards my elaboration of this point from what may seem an unlikely direction, namely, a critique of the fundamental role typically attributed to individual choice in the practice and progress of science. I begin with an autobiographical anecdote.

On a Saturday shortly before Christmas, my father was sent on a Christmas shopping expedition and I, aged about five, was to accompany him. My father did not readily take to the trials and responsibilities of shopping so the ambience was tense. One of father's duties was to buy a present for myself. This purchase was orchestrated by him in the following way. He steered me to a particular toy counter in Woolworth's on which were displayed some six items all priced at two shillings and he invited me to choose. It was with some dismay that I confronted the unattractive options open to me until eventually, under pressure to make a decision, I settled on a somewhat uninspired toy train. We returned home, father's task accomplished and my estimates of the merits of the festive season radically revised. One of the many queries raised by my mother concerning the wisdom of the various purchases focused on the appositeness of my present. 'It was what he chose' was my father's ready reply. My rational faculties were not sufficiently developed to articulate the way in which I had been cheated, but I certainly knew I had been. Perhaps some Oedipal drive came into play from that moment, pushing me towards a career in philosophy. In any event, the moral I would like to draw from the story is this. When it comes to individuals making choices, all the important determinations have already occurred.

Prominent in orthodox philosophy of science is what I regard as an inappropriate emphasis on theory choice. It is typically assumed that the question of why a theory supplants a rival is to be explained in terms of the rational choices of scientists. Theory change is identified with theory choice. I regard this as a misleading and inappropriate identification. There are certainly problems when it comes to formulating what the criteria of theory choice are[1] and no consensus has been reached among those philosophers who have attempted the task. As for scientists themselves, they typically have difficulty understanding the nature of the problem, let alone being able to offer a solution. I suggest that the phenomenon of scientists choosing between rival theories using rational criteria is largely a figment of the analytic philosopher's imagination. Scientists do experiments, they deduce the consequences of theories, compare them with rivals, modify them in the light of problems, and so on. As I have written elsewhere (Chalmers, 1982, p. 116):

> Many scientists contribute in their separate ways with their separate skills to the growth and articulation of physics, just as many workers combine their efforts in the construction of a cathedral. And just as a happy steeplejack may be blissfully unaware of the implication of some ominous discovery made by labourers digging near the foundations, so a lofty theoretician may be unaware of the relevance of some new experimental finding for the theory on which he works.

How does theory change and scientific progress come about as a result of this activity? Elsewhere (Chalmers, 1979; 1980) I introduced the notion of the 'degree of fertility' of a theory to help answer the question. I use the phrase to refer to the extent to which a theory offers opportunities for development in a particular practical or theoretical context, the extent to which a theory opens up lines of development that are real possibilities given the theoretical and experimental resources available. Armed with this conception, we are able to characterize theory change in something like the following way.

Let us suppose that theory A is challenged by theory B. Let us also make the major assumption that there are plenty of scientists with the appropriate skills, resources and frame of mind for working on the rival theories. In such a circumstance, it is highly likely that opportunities for development that in fact exist will sooner or later be taken advantage of. Consequently, if theory B as a matter of fact provides many more opportunities for development than theory A, and provided some of the opportunities bear fruit when taken advantage of, the net effect of the work is that theory B flourishes while theory A stagnates. This construal of scientific practice is not unlike an account of economic development in a (hypothetical) unrestrained capitalist society. Here, while the development is not controlled by any overall rational plan, it is understandable and explicable in terms of objective opportunities for profit-making and the ways in which those opportunities are taken advantage of. Looking at theory change in the way I advocate, we can understand, for example, why Fresnel's version of the wave theory of light had supplanted the particle theory of light by the early 1830s, whereas Young's version of the wave theory had been unsuccessful three decades earlier. Developments in mathematical techniques for dealing with waves in an elastic medium in the early decades of the nineteenth century had the consequence that there were objective opportunities for developing the wave theory that were available to Fresnel but not to Young. To explain the triumph of the wave over the particle theory we do not need to invoke the notion of scientists, armed with rational criteria for theory choice, rationally choosing to remain with the particle theory early in the century but opting for the wave theory by 1830 (Worrall, 1976).

I found a resonance with these largely neglected ideas of mine on theory change in an unexpected quarter, and exploring it will give us an instructive entry into the social and political dimension of scientific practice. The following passage, part of which I utilized in Chapter 6, is taken from an interesting and instructive book, *The Manufacture of Knowledge*, by the sociologist, Karin Knorr-Cetina (1981, p. 8).

> We have heard that validation or acceptance, in practice, is seen as a process of consensus formation qualified as 'rational' by some philosophers, and 'social' by sociologists of science. But whether rational or social, the process appears to be one of opinion formation, and as such, located somewhere else than within scientific investigation itself . . . But where do we find the process of validation, to any significant degree, if not *in* the laboratory itself? If not in the process of laboratory decision-making by which a previous result, a method or a proposed interpretation, comes to be preferred over others and incorporated into new results? What *is* the process of acceptance if not one of selective incorporation of previous results into the ongoing process of research production? To call it a process of

opinion formation seems to provoke a host of erroneous connotations. We do not as yet have science courts for official opinion formation with legislative power in the conduct of future research. To view consensus as the aggregate of individual scientific opinions is misleading, since (a) short of regular opinion polls we have no access to the predominant, general or average opinions of relevant scientists, and (b) it is a commonplace in sociology that opinions have a complex and largely unknown relationship to action. So even if we knew what scientists' opinions were, we would not know which results would be consistently preferred in actual research. What we have, then, is not a process of opinion formation, but one in which certain results are solidified through continued incorporation into ongoing research. This means that the locus of solidification is the process of *scientific investigation*, . . . the *selections* through which research results are constructed *in* the laboratory.

If we equate Knorr-Cetina's usage of 'opinion formation' with my usage of 'rational theory choice' then there is, I suggest, a marked similarity in our views. Where I wish to say that a theory prospers when objective opportunities that if offers for research are taken advantage of, Knorr-Cetina says that a result is solidified to the extent that it is incorporated into ongoing research. However, the ways in which we elaborate our positions are significantly different, and it is by following Knorr-Cetina that we gain a vantage point on the social dimension of scientific practice.

One difference in our respective approaches is that, whereas I have been concerned with macro-theoretical issues such as the replacement of the particle theory by the wave theory of light, Knorr-Cetina focuses on micro-studies of laboratory work. A second difference is that Knorr-Cetina does not take for granted what my account of theory change assumes, for the sake of argument, namely that there will always be scientists with the appropriate skills and resources to take advantage of opportunities for research. The lines of research that are possible in practice for a scientist or group of scientists depend on a range of contingencies such as the availability of the necessary equipment, raw materials, literature, technical assistance and funding. Pursuing the question of how the material and social conditions necessary for research come to be satisfied, either in specific situations or more generally, soon reveals the extent to which scientific practice involves, and cannot be separated from, broader social and political issues.

8.3   The politics of scientific practice

The factors lying behind the satisfaction of the material conditions necessary for scientific work involve a wide range of interests other than the production of scientific knowledge. This point is graphically illustrated by Bruno Latour (1987, pp. 153–7). In a striking passage he compares the day-to-day activity of a scientist in a leading Californian laboratory with the head of the laboratory, whom he refers to as 'the boss'. The scientist sees herself as interested in the development of pure science and as uninterested in politics or broad social issues. She tries to distance herself from government and industry to concentrate on her pure research. By contrast the boss is constantly engaged in political activity at all levels, which frequently earns him the scorn of the scientist.

Latour's example involves research on a new substance, pandorin, which promises to be of great physiological significance. A list of the activities in which the boss is engaged during a typical week includes the following: negotiations with major pharmaceutical companies concerning possible patents of pandorin; a meeting with the French Ministry of Health at which the possibility of opening a new laboratory in France is discussed; a meeting with the National Academy of Science where the boss argues the need for a new subsection; attendance at a board meeting of the journal *Endocrinology* where he urges that more space be given to his area and complains about inadequate referees who know little about the discipline; a visit to the local slaughterhouse where he discusses the possibility of decapitating sheep in a way that will cause less damage to hypothalami; a curriculum meeting at the university where the boss proposes a new curriculum involving more molecular biology and computer science; discussion with a Swedish scientist about his newly devised instruments for detecting peptides and possible strategies for developing it; and an address to the Diabetic Association.

We now follow Latour in turning our attention to the work of the scientist in the laboratory some time later. We find that she has been able to employ a new technician, made possible by a grant from the Diabetic Association, and she has two new graduate students, who have entered the field by way of the new courses designed by the boss. Her research has benefited from the cleaner samples of hypothalami she is receiving from the slaughterhouse and a new, highly sensitive, instrument recently acquired from Sweden which increases her capacity to detect minute traces of pandorin in the brain. Her preliminary results are published in a new section of *Endocrinology*. She is contemplating a position offered to her by the French government involving the setting up of a laboratory in France.

In so far as the scientist in Latour's very believable story regards herself as engaging in pure science untroubled by political and broader social matters she is mistaken. The satisfaction of the material conditions that is a precondition for the conduct of her research is only achieved as a result of political activity involving a range of social interests, as the activities of the boss illustrate. A searching analysis of this aspect of science inevitably leads us to an involvement with broad social and political issues quite distinct from the aim of pure science. If, for example, we probe deeply enough into the source of funds for *any* area of research in physics in the United States we will, more often than not, encounter the interests of the military and the Department of Defense in the development of modern weapons systems. As E. L. Woollett (1980, p. 109) puts it in a revealing article, 'anyone trained in physics who reads the Annual Report of the Secretary of Defense will recognise the essential way in which progress in science has become linked to "progress" in modern weapon systems'. My insistence on the distinction between science and other practices with different aims leaves far more than crumbs for the sociologist to analyse.

The mere fact that scientific practice cannot be separated from other practices serving other interests does not of itself imply that the aim of science is subverted. This much is made apparent by Robert Merton's (1973) somewhat conservative, functionalist analysis of the institutional organization of science. Merton understands science as governed by norms defining the appropriate code of behaviour of scientists, the norms of universalism, disinterestedness, communism and organized scepticism. Allegiance to these norms is presumed to further the aim of science. But individual scientists have their own aims and interests, such as the acquisition of wealth or fame or power. Merton suggests that the aim of science is reconciled with the aims of scientists by way of the institutionalized system of rewards and penalties. In this way scientists are coerced into acting in a way that serves the interests of science because it is precisely that way of acting that results in the rewards that serve their own interests. There are, of course, other interests at stake in the practice of science, such as those of professions, governments and industrial monopolies, and the neglect of these is one of the shortcomings of Merton's analysis. Nevertheless, it does serve to bring out the point that science is not automatically subverted once other interests are involved. We can further illustrate the point by noting that it was a happy coincidence between some aspects of the interests of science and those of the bourgeoisie that enabled science to flourish in the wake of the scientific revolution (see also Bartels and Johnston, 1984).

8.4 Cutting science down to size

I have been concerned in this book to identify and characterize the aim of science and to distinguish it from other practices with different aims. It should not be concluded from this that I regard the aim of science to be some absolute, unqualified good that is necessarily to be rated above other aims. An example will help put an unqualified glorification of science in a more realistic perspective.

In 1815 Humphrey Davy invented the so-called miner's safety lamp. There is no doubt that this resulted from some successful, pure scientific research (possibly carried out by Faraday), involving the determination of the ignition temperature of methane and the effectiveness of a wire gauze in acting as a temperature barrier. J. A. Paris, one of Davy's biographers, referred to this successful research as 'the pride of science, the triumph of humanity and the glory of the age in which we live' (Albury and Schwartz, 1982, p. 13), and, more recently, Union Carbide Chemicals and Plastics, in an advertisement, extol the virtues of Davy's research and liken his contributions to humanity to that of Union Carbide. 'After all, Humphrey Davy lit a lamp on behalf of humanity, and we don't want to see it go out' (Albury and Schwartz, 1982, p. 13). This is not untypical of the way in which the intrinsic value of science is portrayed and glorified.

However, as Albury and Schwartz (1982) point out, a more sober look at the actual history of this episode leads us to a much more qualified evaluation. One immediate effect of the introduction of the Davy lamp into coal-mines was a marked *increase* in the number of explosions and fatalities. The reason for this is not difficult to discern. From the point of view of the mine owners, the pressing problem was not so much mine safety as the fact that coal-rich mine workings rapidly became inaccessible because of the build-up of methane gas. The problem for them, and the one they posed to Davy, was how to get miners into the dangerous, methane-filled workings. Davy's research provided an answer. But, of course, his lamp was far from perfect. The gauze could become detached, draughts could blow the flame outside of the gauze and coal particles settling on its exterior could become red hot. The miners recognized that the most pressing problem in the mines was that of securing adequate ventilation. They realized that the main fatalities after an explosion occurred as a result of suffocation by carbon monoxide and carbon dioxide resulting from the explosion. They proposed measures such as the sinking of additional shafts, but these suggestions were largely ignored, presumably because of the cost involved. The miners could have been forgiven for being sceptical about any claim to the effect that advances in science are an unqualified good.

There are comparable situations today. In the light of adverse effects made possible by science, such as nuclear annihilation or less extreme damage to the environment, it is reasonable to claim, in many contexts, that a more socially equitable use of the scientific knowledge that we have is a more pressing problem than the production of more scientific knowledge. Even when it is appropriate to give a high priority to the acquisition of scientific knowledge, there remains the question of which of the many possible lines of scientific research should be pursued. There remains the question, that is, of what kind of science we want. It is unquestionable that a major driving force underlying the direction of development of Western science stems from the economic and military interests of government agencies and the allied interests of multinational corporations. Many of us wish that things were otherwise, and that science could be developed in directions more in keeping with the needs and interests of ordinary people. In any event, science needs to be evaluated and articulated with references to other interests and values. The evaluations and political struggles involved here are not themselves amenable to scientific solutions.

This latter remark signals the need to grasp the limitations and scope of scientific knowledge. The account of sciences that I have defended construes them as involving specific methods and standards developed in practice to meet specific aims. Once they are understood in this way, it can be appreciated that a wide range of problems lie outside their scope. Even if we restrict the discussion to the behaviour of the physical world, then, if we recall the extent to which support for scientific theories draws on evidence produced under the artificial conditions of a controlled experiment, we can appreciate that complex situations in the real world are beyond the grasp of a complete scientific analysis. For instance, while contemporary science might be quite capable of yielding precise answers to questions concerning the half-life of various components of radioactive waste or the extent to which borosilicate glass disintegrates when exposed to specified degrees of damp, the precise long-term consequences of the probable outcome of various nuclear waste disposal techniques cannot be scientifically determined because our scientific knowledge is not geared to coping with the complexities of real-life situations such as those that obtain when nuclear waste is enclosed in borosilicate glass and buried in deep cavities, or projected into a planetary orbit! While it is important to acknowledge that scientific knowledge is a powerful aid to our technological, engineering and environmental interventions in the world and to our understanding of their possible effects, a recognition of the limitations of science in this respect is a necessary corrective to the mystifications and exaggerations typically accompanying the claims of technocrats (see, for example, Lowe, 1987).

We move further beyond the legitimate domain of science once we introduce questions about the desirability and safety of various technological interventions in the world. Here it is important to avoid obscurantist talk about the interests of humanity in general, which was in evidence in our example involving excessive glorifications of Davy's science, to recognize the variety of interests associated with various individuals, groups and classes and to recognize that those interests frequently conflict. When the safety of a proposed nuclear power station is in question, for example, it makes a great deal of difference from whose point of view the safety is to be assessed, whether it be the owners of the power station, those who are to work in it or live near it, or those industrialists likely to acquire cheap and plentiful electricity as a result of it. The efforts to transform risk analysis into a science, so that the safety of a power station becomes expressed by some objective measure obscure the political conflicts involved, as well as giving a deceptive impression of the precision with which projections are possible.

Much influential, but unfounded, ideology of our time involves an extension of science well beyond its legitimate limits, so that social and political problems are construed as scientific ones and 'solutions' offered in a way that obscures the social and political issues at stake. For example, we have illegitimate extensions of biology and evolutionary theory in the form of social Darwinism and sociobiology posing as explanations of social phenomena, thereby disguising the political realities and serving to justify various kinds of oppression such as that of the poor or women or racial minorities, and in recent times we witness an increasing tendency to reduce social issues to economic ones to be dealt with by a (pseudo)science of economics. It is well beyond the scope of this book to explore such important issues. But a prerequisite for adequately dealing with them is an adequate grasp of the nature of science, of the kinds of achievement it is capable of as well as of its limits.

I am by no means alone in viewing social trends in the contemporary world with dismay and alarm. The gulf between rich and poor and between developed and underdeveloped countries widens, the environment is destroyed and the threat of annihilation looms. The social and political problems facing us are urgent and vital. I do not think this cause is helped by construals of science as a capitalist male conspiracy or as indistinguishable from black magic or voodoo.

And now my nose has started to bleed.

8.5 Note

1 Thomas Kuhn (1977b) identifies some of the problems associated with attempts to construe scientific progress in terms of choices made according to rational criteria, although his response to the problem is quite different from mine.

— APPENDIX

The Extraordinary Prehistory of the Law of Refraction

The law in question is the assertion that, when a ray of light passes from one medium to another, the ratio of the sine of the angle of incidence to the sine of the angle of refraction is a constant characteristic of the pair of media. The law was discovered experimentally by Harriott, derived theoretically and independently by Descartes, and called Snell's law.

Theoretical and experimental studies of reflection date back to Antiquity. The law of reflection was certainly known to Euclid around 300 BC, and early in the second century AD Ptolemy performed experiments to support it. Ptolemy also carried out what seems to have been the first detailed study of refraction. In this article we take up the story, beginning with Ptolemy's work.

The first fascinating feature of the remarkable series of events leading up to the discovery of Snell's law lies in the fact that Ptolemy surreptitiously adjusted his experimental findings so that they conformed to a preconceived idea. Subsequently, following the fall of the Greco-Roman Empire, Arab scientists took up the task of improving on Ptolemy's efforts. In particular, they adopted more sophisticated methods of adjusting the experimental results. Meanwhile, in less mathematically minded Western Europe deceits of a different kind were perpetrated. The extraordinary story nears its end when the two traditions converge on Kepler, who managed to do everything but discover the law of refraction.

Ptolemy's *Optics* is no longer extant. An Arabic version of the original Greek has also been lost. However, a Latin translation of the Arabic version was made in the middle of the twelfth century. This still exists. A readily available English translation of extracts relevant to our topic is in Cohen and Drabkin (1958, pp. 271–81). For the purposes of this article it will be assumed that this English version of the Latin translation of the lost

Arabic rendering of the original Greek corresponds to what Ptolemy actually wrote.

Ptolemy's experiments on refraction differ very little from those that we were all bored by at school. For investigations at an air–water interface the experimental arrangement was as follows. A circular copper disc, with its circumference marked off in one-degree intervals, was supported in a vertical plane with one diameter coinciding with a water surface. A coloured marker was fixed at the centre of the disc, on the air–water interface. A second marker was fixed to the circumference of the disc above the water, so that the line joining the two markers defined an incident ray. A third marker could be moved around the circumference of the disc beneath the water surface until, as viewed from above, it appeared to be in line with the former two markers. The line joining this third marker to the disc's centre then corresponded to the refracted ray. In this way Ptolemy recorded angles of refraction, r, corresponding to angles of incidence, i, ranging from 10° to 80° in ten degree intervals. He performed similar investigations of refraction at air–glass and water–glass interfaces employing a semicylinder of glass.

Ptolemy commented qualitatively on the results. For instance he observed (i) that the incident and refracted rays lie in a plane perpendicular to the refracting surface, (ii) that rays normal to the surface are not refracted, and (iii) that the amount of refraction depends on the densities of the media. He also offered some inequalities. For instance, he pointed out that if i_1 and i_2 are two angles of incidence and if r_1 and r_2 are the corresponding angles of refraction, and if $i_2 > i_1$, then $i_2/i_1 > r_2/r_1$. Ptolemy did not claim that i is proportional to r as some historians, for example A. C. Crombie (1962, p. 120), have claimed.

In addition to his qualitative remarks, Ptolemy presented his numerical results, without commenting on them. The first two columns of Table 1 show these for his investigation of an air–water interface. These results of Ptolemy's possess a regularity. Consecutive values of r differ from each other by an amount that decreases steadily as r increases. The second differences are constant and equal to half a degree. I wish to argue, as others, notably A. Lejeune (1946), have done before, that Ptolemy adjusted his experimental readings so that they possessed this regularity. Restricting myself initially to Ptolemy's work on an air–water interface, I present four arguments that support this charge.

First, since the regularity underlying Ptolemy's results does not correspond to the true state of affairs as expressed by Snell's law, it is most unlikely that the erroneous results would have possessed their regularity by accident. The second argument concerns the discrepancies between the results quoted by Ptolemy and the 'true' values for r shown in the third column of Table 1 calculated from i using a refractive index of 1.33. In the higher and lower regions of the table the discrepancy between Ptolemy's and the correct values are greater than can reasonably be attributed to

Table 1   Ptolemy's experimental results compared with
the correct values

Ptolemy's results		Correct r	
$i°$	$r°$	Calculated for refractive index of 1.33	
10	8	7°	30'
20	15½	14°	54'
30	22½	22°	5'
40	29	28°	54'
50	35	35°	10'
60	40½	40°	37'
70	45½	44°	57'
80	50	47°	46'

experimental error. To establish this I repeated Ptolemy's experiment, reconstructing his apparatus as faithfully as possible following his fairly explicit instructions. Using a scale nine inches in diameter I found that r could be measured to the nearest quarter of a degree quite comfortably, with the exception of the largest angle of refraction where the error could have been as large as one degree. Admittedly, using a scale of three inches in diameter led to much greater probable errors, but there is no reason to believe that Ptolemy would choose to use such an inconveniently small scale. Ptolemy did not specify the dimensions of his apparatus.

Third, there is evidence that Ptolemy did indeed believe that there was a definite relationship between i and r. For he claimed to have shown 'that this type of bending (refraction) does not take place at equal angles but that the angles, as measured from the perpendicular, have a definite quantitative relationship' (Cohen and Drabkin, 1958, p. 272). Ptolemy offered no evidence to support this claim, unless the order underlying his results is taken to constitute such evidence.

The fourth point lending plausibility to the claim that Ptolemy adjusted his readings so that the second differences were constant is the fact that Babylonian astronomical tables portraying the angular distance traced out by the sun in successive months, and almost certainly known to Ptolemy, possessed just that feature. Mathematical series very familiar to the Ancients, such as the sequence of the squares of the natural numbers, also possess it. It is worth noting that at that stage in history the notion of a continuous mathematical *function* linking one variable with another had yet to be developed. If there was to be a 'definite quantitative relationship' connecting i and r, then tables of the kind we are discussing constituted the only mathematical tools available to Ptolemy for expressing such a relationship.

The fact that Ptolemy's recorded values for *r* differ from the correct values more at the two extremes of the table than they do in the middle suggests that Ptolemy started from the middle of the table of his measured values, which we can assume were within half a degree of the correct values, and adjusted the values at the two extremes until the second differences were constant and equal to half a degree. He was then prepared to allow the adjusted readings to pose as his recorded ones, according to that well-tested method very familiar to science students.

So far the case against Ptolemy has referred only to his results for an air–water surface. The case is much strengthened by the observation that Ptolemy's other two tables of results, for air–glass and water–glass interfaces, show precisely the same regularity. The second differences are again constant and equal to half a degree.

In the light of the foregoing, the reference by G. Sarton (1927, p. 268) to Ptolemy's work in optics as 'the most remarkable experimental investigation of antiquity' and the remark by B. Farrington (1963, p. 294) that 'we observe here, as elsewhere, a combination of insight and system charactertistic of the man' take on a somewhat ironic aspect.

Optical investigations by early Arabic scientists were the first sequel to Ptolemy's work. Those scientists were familiar with Ptolemy's *Optics* and attempted to improve on it in a variety of ways. Alhazen (AD 965–1039) wrote one major and several minor works on optics. He tried to improve on Ptolemy's experimental findings by constructing a more elaborate piece of apparatus, the main improvement being the substitution of a narrow ray of light from the sun or a candle for Ptolemy's markers. Alhazen came to recognize that Ptolemy's results were not accurate, although he still employed them when attempting to work out the focusing properties of a glass sphere. He did not quote any of his own measurements of *r*.

More significant for our story is the work performed three centuries later by al-Farisi. Al-Farisi recognized the order underlying Ptolemy's results and endeavoured to improve on them, not by performing experiments but by employing 'improved' methods of calculation. Using Ptolemy's readings of *r* corresponding to $i = 40°$ and $i = 50°$ at an air–glass interface, al-Farisi explicitly used a 'refined' method, employing series with constant first, second and third differences, to calculate *r* for values of *i* ranging from 1° to 17° in one-degree intervals. The resulting values of *r* differ more from the correct values than Ptolemy's did. The techniques employed by al-Farisi were common among Arab astronomers and had their origin in the Babylonian astronomy mentioned above. In relating one variable to another using tables of increasing complexity the Arabic scientists were moving ever closer to the idea of a continuous function (for details see Schramm, 1965).

By contrast to these early Arab scientists, their Western European contemporaries were much less mathematically sophisticated, adopting a more qualitative approach under the influence of the writings of Aristotle. Medieval Western European writers frequently referred to the importance of experiments in science, but the claim of Crombie (1962) to the effect that in their work can be seen the origins of the modern experimental method is hardly borne out by their work on refraction. According to Crombie, Robert Grosseteste was one of the early pioneers of the experimental method. Grosseteste claimed that reflection and refraction of light can best be studied experimentally. However, the law of refraction that he proposed, namely, that the angle of refraction is half the angle of incidence, can be very simply refuted by experiment. This discrepancy, between the lip-service paid to experiment, on the one hand, and the lack of concrete results achieved by actually experimenting, on the other, is even more marked in the writings of the Silesian scientist, Witelo, to a discussion of whose work we now turn.

In about 1270 Witelo wrote a textbook on optics drawing on all the sources available to him, including the works of Ptolemy and Alhazen. This was to become the standard work in optics for no less than three and a half centuries, not only because of the comprehensiveness of its subject matter but also because, unlike its predecessors, it was written in readable Latin. In his book Witelo discussed refraction. One of his assertions about that topic is followed by this passage (Crombie, 1962, p. 219):

> Proof of this proposition depends on experiments with instruments rather than on other kinds of demonstration. Therefore when someone wants to find out the manner in which rays of light are refracted in a second transparent medium denser than the first, as in water, which is denser than air (let him use the instrument described by Alhazen) . . .

Witelo went on to describe in great detail an apparatus for measuring angles of refraction which was an improved version of Alhazen's. With his apparatus he could measure r for rays passing in either direction across air–water, air–glass and water–glass interfaces. His alleged results for air–water are shown in Table 2. The differences between i and r were included in Witelo's tables. The error in subtraction in the first row is Witelo's.

Let us look at the results for air to water first. With the exception of the first reading these are identical to Ptolemy's and Witelo has presumably copied them. His results for air to glass are identical to Ptolemy's without exception. The fact that Witelo was prepared to change the first reading has great significance. Witelo's modification destroys the order underlying the readings and indicates that he was not aware of that order. He was ignorant of mathematical techniques very familiar to his Arab contemporaries.

We now turn to the second half of Witelo's table. Since rays passing from water to air experience total internal reflection for large angles of

Table 2 Witelo's 'Experimental' results

$i°$	Air–water				Water–air			
	r		$i - r$		r		$r - i$	
10	7°	45′	2°	5′	12°	5′	2°	5′
20	15°	30′	4°	30′	24°	30′	4°	30′
30	22°	30′	7°	30′	37°	30′	7°	30′
40	29°	0′	11°	0′	51°	0′	11°	0′
50	35°	0′	15°	0′	65°	0′	15°	0′
60	40°	30′	19°	30′	79°	30′	19°	30′
70	45°	30′	24°	30′	94°	30′	24°	30′
80	50°	0′	30°	0′	110°	0′	30°	0′

incidence, one glance from a modern eye indicates that the results are absurd, and could not have resulted from experimental measurements. It is not difficult to see how Witelo has *calculated* these angles of refraction. His calculation was apparently based on a misinterpretation of Ptolemy who, in his *Optics*, wrote

> Our proposition is that the *amount* of the refraction is the same in each of the two types of passage but that the two refractions differ in type. For in its passage from a rarer to a denser medium the ray inclines *toward* the perpendicular, whereas in its passage from a denser to a rarer medium it inclines *away* from the perpendicular (Cohen and Drabkin, 1958, p. 279).

This is presumably a rather careless statement of the law of reversibility. Witelo has interpreted it to mean that if, for some angle of incidence, a ray passing from air to water is deflected by x degrees towards the normal, then a ray passing from water to air at the same angle of incidence will be deflected by x degrees away from the normal.

The fact that Witelo's readings are half copied and half calculated from a false theory and that the second set in particular bear little relation to what actually happens detracts considerably from the credibility and significance of Witelo's sermons on the importance of experiment.

During the three centuries following the events just described more and more Greek and Arabic texts became known to Western Europeans. Rennaissance scientists were much more mathematically sophisticated than their predecessors, and less Aristotelian. In the first decade of the seventeenth century the next person of note in our story, Kepler, turned his considerable theoretical skills to a study of optics. His initial source was Witelo's text. When he saw what he regarded as Witelo's results on refraction, but which we know were in fact Ptolemy's, his familiarity with the techniques of astronomers was sufficient for him to see immediately the order behind the readings. However, Kepler disagreed that the relationship between i and r should be of that kind. He was convinced that the correct relationship would take the form of a *trigonometric function*. Kepler tested a variety of trigonometric functions against experiment. He tried $i - r = k \sec i$, $2i - r = k \sin i$, $\tan i = k \tan r$, $\tan i = \sin(i - r)$, $1 - \tan i \cot (i - r) = k \tan i$ (here he begins to get desperate), $1 - \tan i \cot (i - r) = k \sin i$, $i - r = k_1 + k_2 \sec i$, and finally, $1 - \tan i \cot (i - r) = k_1 + k_2 \sin i$. But none of these proved satisfactory and at this stage the unfortunate Kepler conceded defeat.

Discovery of the correct law was now not far away. It would seem that there were eventually at least three independent discoverers. Manuscripts in the British Museum discussed by Shirley (1951) indicate that Thomas Harriott discovered the law experimentally around 1616, although he did not publicize it. Snell also discovered the law some time before 1626 for it is mentioned in his manuscripts. He, too, did not publish his result and it is not clear how he arrived at it. Descartes derived the sine law theoretically, possibly as early as 1619, and was certainly the first to publish it, in 1637. Sabra (1967, ch. 4) has argued persuasively that Descartes arrived at the law independently of and possibly before Snell, contrary to what others have claimed. To a modern physicist Descartes' derivation will be most unconvincing since it is based on false assumptions and the argument is not entirely conclusive. However, Sabra has shown that Descartes' argument makes a great deal of sense when his theory is viewed in its historical context.

The history of the law of refraction certainly strikes a blow at any simple-minded idea of science progressing steadily by careful generalizations from the results of observation and experiment. The antics of the early investigators serve to show that the experimental method that modern science takes for granted did not always exist. The craft skills involved in experimentation, the careful elimination of sources of error, the repetition and criticism of readings, the estimation of probable errors and so on, gradually came into being during the seventeenth century owing much to pioneers such as Kepler and Galileo. Modern scientists are much more skilled than Ptolemy was at knocking nature into shape.

Bibliography

Albury, D. and Schwartz, J. (1982). *Partial Progress: The Politics of Science and Technology*. London, Pluto Press.

Albury, R. (1983). *The Politics of Objectivity*. Victoria, Deakin University Press.

Althusser, L. (1966). *For Marx*. Harmondsworth, Penguin.

Anscombe, E. and Geach, P. T. (eds) (1977). *Descartes: Philosophical Writings*. Milton Keynes, Open University Press.

Armstrong, D. (1973). *Belief, Truth and Knowledge*. Cambridge, Cambridge University Press.

Barnes, B. (1977). *Interests and the Growth of Knowledge*. London, Routledge and Kegan Paul.

Barnes, B. and Mackenzie, D. (1979). 'On The Role of Interests in Scientific Change' in R. Wallis (ed.), *On the Margins of Science: The Social Construction of Rejected Knowledge*. Keele, University of Keele Press, pp. 139–78.

Barnes, B. and Bloor, D. (1982). 'Relativism, Rationalism and The Sociology of Knowledge' in M. Hollis and S. Lukes (eds), *Rationality and Relativism*. Oxford, Basil Blackwell.

Barnes, J. (1975). 'Aristotle's Theory of Demonstration' in J. Barnes, M. Schofield and R. Sorabji (eds), *Articles on Aristotle, I: Science*. London, Duckworth.

Bartels, D. and Johnston, R. (1984). 'The Sociology of Goal-Directed Science: Recombinant DNA Research', *Metascience*, 1/2, 37–45.

Bhaskar, R. (1978). *A Realist Theory of Science*. Hassocks, Harvester.

Block, I. (1961). 'Truth and Error in Aristotle's Theory of Perception', *Philosophical Quarterly*, 11, 1–9.

Bloor, D. (1976). *Knowledge and Social Imagery*. London, Routledge and Kegan Paul.

Bloor, D. (1981). 'The Strengths of the Strong Programme', *Philosophy of Social Science*, 11, 173–98.

Bloor, D. (1982). 'Durkheim and Mauss Revisited: Classification and the Sociology of Knowledge', *Studies in History and Philosophy of Science*, 13, 267–97.

Bock, J. W. (1973). 'Philosophical Foundations of Classical Evolutionary Classification', *Systematic Zoology*, 22, 375–92.

Burtt, E. A. (1967). *The English Philosophers from Bacon to Mill*. New York, Random House.

Chalmers, A. F. (1973). 'The Limitations of Maxwell's Electromagnetic Theory', *Isis*, 64, 469–83.

Chalmers, A. F. (1975). 'The Extraordinary Prehistory of the Law of Refraction', *The Australian Physicist*, 12, 85–8.

Chalmers, A. F. (1979). 'Towards an Objectivist Account of Theory Change', *British Journal for the Philosophy of Science*, 30, 227–33.

Chalmers, A. F. (1980). 'An Improvement and a Critique of Lakatos's Methodology of Scientific Research Programmes', *Methodology and Science*, 13, 2–27.

Chalmers, A. F. (1982). *What Is This Thing Called Science?*, 2nd revised edn. Milton Keynes, Open University Press.

Chalmers, A. F. (1984). 'A Non-Empiricist Account of Experiment', *Methodology and Science*, 17, 95–114.

Chalmers, A. F. (1985a). 'The Case against a Universal Ahistorical Scientific Method', *Bulletin of Science, Technology and Society*, 5, 555–67.

Chalmers, A. F. (1985b). 'Galileo's Telescopic Observations of Venus and Mars', *British Journal for the Philosophy of Science*, 36, 175–91.

Chalmers, A. F. (1986). 'The Galileo that Feyerabend Missed: An Improved Case Against Method' in J. A. Schuster and R. A. Yeo (eds), *The Politics and Rhetoric of Scientific Method*. Dordrecht, Reidel, pp. 1–31.

Chalmers, A. F. (1988). 'The Sociology of Knowledge and the Epistemological Status of Science', *Thesis Eleven*, 21, 82–102.

Clavelin, M. (1974). *The Natural Philosophy of Galileo*. Cambridge, Mass., MIT Press.

Cohen, M. R. and Drabkin, I. E. (1958). *A Source Book in Greek Science*. Cambridge, Mass., Harvard University Press, pp. 271–81.

Collier, A. (1979). 'In Defence of Epistemology' in J. Mepham and D.-H. Ruben (eds), *Issues in Marxist Philosophy, Vol. 3: Epistemology, Science, Ideology*. Hassocks, Harvester, pp. 55–106.

Collins, H. M. (1981). 'Son of Seven Sexes: The Social Destruction of a Physical Phenomenon', *Social Studies of Science*, 9, 33–62.

Collins, H. M. (1983). 'An Empirical Relativist Programme in the Sociology of Scientific Knowledge' in Knorr-Cetina and Mulkay (1983, pp. 85–113).

Collins, H. M. (1985). *Changing Order: Replication and Induction in Scientific Practice*. London, Sage.

Collins, H. M. and Cox, G. (1976).'Recovering Relativity: Did Prophecy Fail?', *Social Studies of Science*, 6, 423–44.

Crombie, A. C. (1962). *Robert Grosseteste and the Origins of Experimental Science*. Oxford, Oxford University Press.

Drake, S. (1957). *Discoveries and Opinions of Galileo*. New York, Double-day Anchor.

Drake, S. (1973). 'Galileo's Experimental Confirmation of Horizontal Inertia', *Isis*, 64, 291–305.

Drake, S. (1978). *Galileo at Work*. Chicago, University of Chicago Press.

Drake, S. (1983). *Telescopes, Tides and Tactics*. Chicago, University of Chicago Press.

Edge, D. O. and Mulkay, M. J. (1976). *Astronomy Transformed*. New York, Wiley Interscience.

Farrington, B . (1963). *Greek Science*. London, Penguin.

Feyerabend, P. K. (1975). *Against Method*. London, New Left Books.

Feyerabend, P. K. (1976). 'On the Critique of Scientific Reason' in Howson (1976, pp. 209–39).

Feyerabend, P. K. (1981). 'On the Interpretation of Scientific Theories' in P. K. Feyerabend, *Realism, Rationalism and Scientific Method: Philosophical Papers, Vol. 1*. Cambridge, Cambridge University Press, pp. 37–43.

Feyerabend, P. K. (1987). *Farewell to Reason*. London, Verso.

Freudenthal, G. (1986). *Atom and Individual in the Age of Newton*. Dordrecht, Reidel.

Gaffney, E. S. (1979). 'An Introduction to the Logic of Phylogeny Reconstruction' in J. Cracraft and N. Eldredge (eds), *Phylogenetic Analysis and Paleontology*. New York, Columbia University Press, pp. 79–111.

Galileo (1957). *The Starry Messenger* in Drake (1957).

Galileo (1960). *On Motion and On Mechanics*, transl. S. Drake. Madison, University of Wisconsin Press.

Galileo (1967). *Dialogue Concerning the Two Chief World Systems*, transl. S. Drake. Berkeley, University of California Press.

Galileo (1974). *Two New Sciences*, transl. S. Drake. Madison, University of Wisconsin Press.

Gaukroger, S. (1978). *Explanatory Structures*. Hassocks, Harvester.

Geymonat, L. (1965). *Galileo Galilei*. New York, McGraw-Hill.

Gower, B. (1988). 'Chalmers on Method', *British Journal for the Philosophy of Science*, 39, 59–65.

Hacking, I. (1983). *Representing and Intervening*. Cambridge, Cambridge University Press.

Hanfling, O. (1981). *Logical Positivism*. Oxford, Basil Blackwell.

Hanson, N. R. (1958). *Patterns of Discovery*. Cambridge, Cambridge University Press.

Hanson, N. R. (1969). *Perception and Discovery*. San Francisco, Freeman and Cooper.

Hertz, H. (1962). *Electric Waves*. New York, Dover.

Hiebert, E. (1988). 'The Role of Experiment and Theory in the Development of Nuclear Physics' in D. Batens and J. P. Van Bendegem (eds), *Theory and Experiment: Recent Insights and New Perspectives on Their Relation*. Dordrecht, Reidel.

Hon, G. (1987). 'H. Hertz: "The electrostatic and electromagnetic properties of the cathode rays are either *nil* or very feeble" (1883). A Case-study of an Experimental Error', *Studies in History and Philosophy of Science*, 18, 367–82.

Howson, C. (ed.) (1976). *Method and Appraisal in the Physical Sciences*. Cambridge, Cambridge University Press.

Hume, D. (1969). *A Treatise on Human Nature*. Harmondsworth, Penguin.

Jacob, M. C. (1976). *The Newtonians and the English Revolution 1689–1720*. Ithaca, Cornell University Press.

Knorr-Cetina, K. D. (1981). *The Manufacture of Knowledge*. Oxford, Pergamon Press.

Knorr-Cetina, K. D. (1983). 'The Ethnographic Study of Scientific Work: Towards a Constructivist Interpretation of Science' in Knorr-Cetina and Mulkay (1983, pp. 115–40).

Knorr-Cetina, K. D. and Mulkay, M. (eds) (1983). *Science Observed: Perspective in the Social Study of Science*. London, Sage.

Koertge, N. (1977). Galileo and the Problem of Accidents, *Journal of the History of Ideas*, 38, 389–408.

Kuhn, T. S. (1959). *The Copernican Revolution*. New York, Random House.

Kuhn, T. S. (1970). *The Structure of Scientific Revolutions*. Chicago, University of Chicago Press.

Kuhn, T. S. (1977a). 'Mathematical versus Experimental Traditions in the Development of Physical Science' in T. S. Kuhn, *The Essential Tension*. Chicago, University of Chicago Press, pp. 31–65.

Kuhn, T. S. (1977b). 'Objectivity, Value-Judgement and Theory Choice' in T. S. Kuhn, *The Essential Tension*. Chicago, University of Chicago Press, pp. 320–39.

Lakatos, I. (1968). Changes in the Problem of Inductive Logic in I. Lakatos (ed.), *The Problem of Inductive Logic*. Amsterdam, North Holland; reprinted in Worrall and Currie (1978b).

Lakatos, I. (1974). 'Falsification and the Methodology of Scientific Research Programmes' in Lakatos and Musgrave (1974, pp. 91–196); reprinted in Worrall and Currie (1978a).

Lakatos, I. (1978a). 'History of Science and its Rational Reconstructions' in Worrall and Currie (1978a, pp. 102–38).

Lakatos, I. (1978b). 'Newton's Effect on Scientific Standards' in Worrall and Currie (1978a, pp. 193–222).

Lakatos, I. (1978c). 'Science and Pseudo-Science' in Worrall and Currie (1978a, 1–7).

Lakatos, I. and Musgrave, A. (eds) (1974). *Criticism and the Growth of Knowledge*. Cambridge, Cambridge University Press.

Latour, B. (1987). *Science in Action*. Milton Keynes, Open University Press.

Latour, B. and Woolgar, S. (1979). *Laboratory Life: The Social Construction of Scientific Facts*. London, Sage.

Laudan, L. (1977). *Progress and its Problems: Towards a Theory of Scientific Growth*. London, Routledge and Kegan Paul.

Laudan, L. (1981). 'The Pseudo-Science of Science?', *Philosophy of Social Science*, 11, 173–98.

Laudan, L. (1984). *Science and Values: The Aims of Science and Their Role in Scientific Debate*. Berkeley, University of California Press.

Lejeune, A. (1946). Les Tables de Refractions de Ptolémée. *Annales de la Société Scientifique de Bruxelles*, 60, 13–101.

Locke, J. (1967). *An Essay Concerning Human Understanding*. London, Dent.

Lowe, I. (1987). 'Measurement and Objectivity: Some Problems of Energy Technology' in J. Forge (ed.) *Measurement, Realism and Objectivity*. Dordrecht, Reidel.

Mackenzie, D. (1978). Statistical Theory and Social Interests: A Case Study. *Social Studies of Science*, 8, 35–83.

Mackenzie, D. (1981). *Statistics in Britain: 1865–1930*. Edinburgh, Edinburgh University Press.

Maxwell, J. C. (1965). 'Illustrations of the Dynamical Theory of Gases' in W. D. Niven (ed.) *The Scientific Papers of James Clerk Maxwell*. New York, Dover, Vol. 1, pp. 377–409.

Merton, R. K. (1973). *The Sociology of Science*. Chicago, University of Chicago Press.

Mulkay, M. (1979). *Science and the Sociology of Knowledge*. London, Allen & Unwin.

Musgrave, A. (1974a). 'The Objectivism of Popper's Epistemology' in P. A. Schilpp (ed.), *The Philosophy of Karl Popper*. La Salle, Ill., Open Court, pp. 560–96.

Musgrave, A. (1974b). 'Logical versus Historical Theories of Confirmation', *British Journal for the Philosophy of Science*, 25, 1–23.

Nickles, T. (1987). 'Lakatosian Heuristics and Epistemic Support', *British Journal for the Philosophy of Science*, 38, 181–205.

Pickering, A. (1981). The Hunting of the Quark, *Isis*, 72, 216–36.

Popper, K. R. (1961). *The Poverty of Historicism*. London, Routledge and Kegan Paul.

Popper, K. R. (1966). *The Open Society and its Enemies*, Vol. 2. London, Routledge and Kegan Paul.

Popper, K. R. (1972). *The Logic of Scientific Discovery*. London, Hutchinson.

Popper, K. R. (1974). 'Normal Science and its Dangers' in Lakatos and Musgrave (1974).

Popper, K. R. (1979). *Objective Knowledge*. Oxford, Oxford University Press.

Popper, K. R. (1983). *Realism and the Aim of Science*. London, Hutchinson.

Porter, T. M. (1981). 'A Statistical Survey of Gases: Maxwell's Social Physics', *Historical Studies in the Physical Sciences*, 12, 77–116.

Price, D. J. de S. (1969). 'A Critical Re-estimation of the Mathematical Planetary Theory of Ptolemy' in M. Clagett (ed.), *Critical Problems in the History of Science*. Madison, University of Wisconsin Press, pp. 197–218.

Rorty, R. (1980). *Philosophy and the Mirror of Nature*. Oxford, Blackwell.

Sabra, A. J. (1967). *Theories of Light from Descartes to Newton*. London, Oldbourne.

Sarton, G. (1927). *Introduction to the History of Science*, Vol. 1. Baltimore, MD, Williams and Wilkins.

Schramm, M. (1965). 'Steps Towards the Idea of Function: A Comparison Between Eastern and Western Science of the Middle Ages', *History of Science*, 4, 70–103.

Shapin, S. (1982). 'History of Science and its Sociological Reconstructions', *History of Science*, 20, 157–211.

Shea, W. R. (1972). *Galileo's Intellectual Revolution*. London, Macmillan.

Shirley, J. R. (1951). 'An Early Experimental Determination of Snell's Law', *American Journal of Physics*, 19, 307–8.

Suchting, W. (1983). 'Knowledge and Practice: Towards a Marxist Critique of Traditional Epistemology', *Science and Society*, 44, 2–36.

Theocharis, T. and Psimopoulos, M. (1987). 'Where Science Has Gone Wrong', *Nature*, 329, 15 October, 595–8.

Thomson, W. and Tait, P. G. (1879). *Treatise on Natural Philosophy*. Cambridge, Cambridge University Press.

Thurber, J. (1933). *My Life and Hard Times*. New York, Harper and Brothers.

Tiles, M. (1984). *Bachelard: Science and Objectivity*. Cambridge, Cambridge University Press.

Turnbull, D. (1984). 'Relativism, Reflexivity and the Sociology of Scientific Knowledge', *Metascience*, 1/2, 47–60.

Wallace, W. (1974). 'Theodoric of Freiberg: On the Rainbow' in E. Grant (ed.), *A Source Book in Medieval Science*. Cambridge, Mass., Harvard University Press, pp. 435–41.

Wallace, W. (1981). *Prelude to Galileo*. Dordrecht, Reidel.

Watkins, J. (1985). *Science and Scepticism*. Princeton, NJ, Princeton University Press.

Wisan, W. L. (1978). 'Galileo's Scientific Method: A Reconstruction' in R. E. Butts and J. C. Pitt (eds), *New Perspectives on Galileo*. Dordrecht, Reidel, pp. 1–57.

Woolgar, S. (1981). 'Interests and Explanation in the Social Study of Science', *Social Studies of Science*, 11, 365–94.

Woollett, E. L. (1980). Physics and Modern Warfare: The Awkward Silence. *American Journal of Physics*, 48, 2, 104–11.

Worrall, J. (1976). 'Thomas Young and the "Refutation" of Newtonian Optics: A Case Study in the Interaction of Philosophy of Science and History' in Howson (1976, pp. 107–79).

Worrall, J. (1988). 'The Value of a Fixed Methodology', *British Journal for the Philosophy of Science*, 39, 263–75.

Worrall, J. and Currie, G. (eds) (1978a). *Imre Lakatos, Philosophical Papers, Volume 1: The Methodology of Scientific Research Programmes*. Cambridge, Cambridge University Press.

Worrall, J. and Currie, G. (eds) (1978b). *Imre Lakatos. Philosophical Papers, Volume 2: Mathematics, Science and Epistemology*. Cambridge, Cambridge University Press.

Yearley, S. (1982). 'The Relationship between Epistemological and Sociological Cognitive Interests: Some Ambiguities Underlying the Use of Interest Theory in the Study of Scientific Knowledge', *Studies in History and Philosophy of Science*, 13, 353–88.

Young, R. (1969). 'Malthus and the Evolutionists: The Common Context of Biological and Social Theory', *Past and Present*, 43, 109–41.

Young, R. (1971). 'Darwin's Metaphor: Does Nature Select?', *The Monist*, 55, 442–503.

Author Index